Dear Mr. Logsdon,

I have just finished reading [illegible] book, *At Nature's Pace*, again. . . . I turn to it much as one would to an old friend, for that is what it has become since I first discovered you on a bookshelf in Pittsburgh, Pennsylvania, in early 1993. This book has since been joined by others penned by you, earlier and later. From all I derive the same freshness, honesty, and passion.

I write this letter as a thank you, since you articulate what I could not, and you have given voice to the ruminations of my mind over the past decade. The elegant blend of a small, diversified farm supplemented with off-farm income awaits implementation in twenty-two months, when the United States Army no longer controls my whereabouts.

Having grown up in Lancaster County, Pennsylvania, I always gave great respect to those involved with farming and the farming community, and it wasn't until I purchased your book in 1993 that I encountered opinion to the contrary. While I read your book in the student lounge at the University of Pittsburgh law school, several acquaintances guffawed at the notion that the American Dream could involve something as "simple" and "unskilled" and "mundane" as farming. That summer I was married, and as wedding favors my wife and I gave white oak seedlings, sprouted from the acorns dropped by the trees in front of the law school. Those who had earlier mocked my reading material thanked me at the reception for the pretty seedlings. I just smiled.

The Army has sent me to Missouri, and now to New York. I never realized how good the soil in Lancaster was, until I had to contend with what passes for soil in the Ozarks and West Point. It also saddens me to see Lancaster slowly paved over, as some of the finest non-irrigated farmland in the world is lost to urban sprawl. I wonder if I can return home. A mall and retirement community now sit on my best friend's cornfield. The stars of the night sky are dimmed by the competition from ambient lighting on the ground. In two years my family will need to decide where to put down roots, as wanting to be rooted to a place is the primary reason I am leaving the service. Your chapter, "Knowing One's Place," reminded me of what it is I want to do with the time I am allotted on this earth. As we begin our search, I thought I should revisit the book that caused me to begin the journey I am currently on. It was reassuring to finish the book with the same feelings I had when I first read it.

As my two daughters grow I want them to know of their connections to both the ground they walk on and the community that surrounds them. I want to participate in their growth, and provide them with more than the ability to accumulate possessions. And I am pleased that I have found voices to which I can relate. First you, then Wendell Berry, Marty Strange, Noel Perrin, and Donald Hall, to name a few. I am not one to write letters, let alone to folks I have never met, but I felt compelled to write. I cannot express my level of gratitude.

Sincerely,
Jeff Hatch

Living at Nature's Pace

FARMING AND THE AMERICAN DREAM

Gene Logsdon

CHELSEA GREEN PUBLISHING COMPANY
White River Junction, Vermont

Dedicated to Andy Reinhart and Jan Dawson, the kind of farmers who will save us yet, for their tremendous example and support.

Designed by Cheryl L. Cipriani and Suzanne Church.

Printed in the United States on recycled, acid-free paper.

03 02 2 3 4 5

Originally published in 1993 by Pantheon Books, a division of Random House. Revised edition published by Chelsea Green in February, 2000.

The following essays were originally published in various publications as indicated: "Green Fields, Red Ink": *Ohio Magazine* (April 1986); "For Amber Waves of Green": *Ohio Magazine* (March 1980); "Our Hidden Wound": *Farm Journal* (March 1992); "The Failure of Agricultural Education": *Ohio Magazine* (May 1992); "Traditional Farming" from *Meeting the Expectations of the Land*, ed. by Wes Jackson, Wendell Berry, Bruce Colman (North Point Press, 1984); "Knowing One's Place": *Draft Horse Journal* (Spring 1982); "The Future: More Farmers, Not Fewer": *Whole Earth Review* (Spring 1989); "Rural Rides Revisited," published in slightly different form in *Draft Horse Journal* (Winter 1989-90); "Amish Economics": *Whole Earth Review* (Spring 1986) and *Orion* (Spring 1988); "A Horse-Drawn Economy": *Draft Horse Journal* (Winter 1988-89); "The Barn Raising": *Ohio* (May 1983); "Not So Friendly Persuasion": *Ohio* (October 1983); "A Patriarch Passes": *Ohio* (July 1988); "A Woodcutter's Pleasures": *Organic Gardening* (October 1988); "The Pond at the Center of the Universe": *Ohio* (June 1991); "My Wilderness": *Organic Gardening* (February 1980); "The Wheel of Life Turns Round and Round," published in a slightly different form in *The Land Report*, ed. by Wes Jackson (The Land Institute, Spring 1999); "Going to Market on a Warm Day in November": *Country Journal* (November/December 1999); "The Folly of Trying to Repress the Agrarian Impulse" from *The Land Report*, ed. by Wes Jackson (The Land Institute, Fall, 1999).

Library of Congress Cataloging-in-Publication Data

Logsdon, Gene.
Living at nature's pace : farming and the American dream / Gene Logsdon.-- 1st Chelsea Green ed.
p. cm.
Originally published: New York : Pantheon Books, 1994.
ISBN 1-890132-56-X (alk. paper)
1. Agriculture--United States. 2. Amish--United States. 3. Country life--United States. I. Title.
S441 .L613 2000
630'.973--dc21 99-057305

Chelsea Green Publishing Company
Post Office Box 428
White River Junction, VT 05001
(800) 639-4099
www.chelseagreen.com

Contents

Foreword

This is a book for which I have been waiting for a long time. Gene Logsdon is my friend; I don't know how I would have got along without him. But I don't think I am speaking from prejudice when I say that he is the best agricultural writer we have.

One reason he is the best is that he is superbly qualified. He was born and bred a farmer, in the circumstances he describes in the first of these essays ("Green Fields, Red Ink"). Later, he left his home country in Wyandot County, Ohio, experienced what to him was enough of the life of university and city and suburb, and then returned home again. Since his return, he and his wife, Carol, have turned twenty acres into the best, the most productive, and the most beautiful small farm I have ever seen. Going to the Logsdons' when the raspberries are ripe, you feel you are getting a little glimpse of Eden—but of course that is a visitor's illusion; the place is excellent because it is the product of excellent work, and of much sweat.

Over the years, Gene has written for a magazine expounding the conventional industrial agricultural doctrine, and then for another magazine expounding the conventional organic agricultural doctrine, and so he knows the issues and the arguments from both sides. By the same token, he knows that between doctrine and practice there is a finally irreducible difference, for farming is an encounter, not with an idea but with a place. He has visited hundreds of farms and talked with hundreds of farmers. He understands thoroughly the practices and the problems of farming. He knows his subject from study, from observation, and from experience.

But Gene's best qualification is that he loves farming. Most people who have influenced and written about farming in our time have not loved it at all; they have held it in contempt. That is why they have tried to turn farms into factories and farmers into factory workers—and, in the process, have nearly destroyed both. Gene loves farming, I think, because he knows that a real farm can never be a factory, and that a real farmer has not the faintest resemblance to a factory worker. He knows that real farming involves a fundamental capacity to be excited about the life of field and pasture and woodlot, and a fundamental affection for animals and for one's neighborhood.

To love farming—real farming—in this day and time requires what a lot of people like to call crankiness but is in fact courage. I have been reading Gene Logsdon for many years, and I have always taken courage from him. I thank him, and I shake his hand.

—*Wendell Berry*

Acknowledgments

I wish there were space to thank all the people who have contributed spiritually or materially to this book, but I must mention at least: Andy Reinhart and Jan Dawson, Wendell and Tanya Berry, Maury and Jeannine Telleen, Kamyar Enshayan and Laura Jackson, Wes Jackson, Jerry and Jill Carlson, Dave Frey, Dave Smith at Seeds of Change, David and Elsie Kline, Steve and Pat Gamby, and Seth Wilpan. For their work in encouraging and editing this new edition, my gratitude to the folks at Chelsea Green, especially Jim Schley, Rachael Cohen, and Hannah Silverstein. I also want to thank those economists at Ohio State University and elsewhere who, with their naïve support of monopolistic tendencies in agriculture, have pissed me off enough to bring this book to completion.

Introduction

In the introduction to the first edition of *At Nature's Pace*, I said that I wrote this book out of anger. That declaration still stands, although the anger, as expressed in the new essays, is tempered with patience and sorrow. Or to put it another way, I haven't mellowed, I've just gotten sneakier. I remain, to the end, unreconstructed.

I am still angry about the decline of rural society, which leads unerringly to a decline in urban society. I still think it doesn't have to happen. I still refuse to bow to the "wisdom" of accepting this decline simply because I am powerless to prevent it or because smart people are sure it is inevitable. I still don't believe in the economist's god of inevitable economic determinism. I maintain that human willfulness can save society, or can destroy a civilization by actively choosing short-term gain over long-term good or by passively remaining silent and supine in the face of someone else's choice of short-term gain. As I say in this book, it surpasses my understanding how we can agree to regulate players in sports to act properly for the long-term survival of our stupid games, but we cannot do the same with real life and real community.

I'm still angry at our educational system, which could teach the folly of allowing a rural society to languish and die, but instead perpetrates the myth that bigger urban schools are better than smaller neighborhood schools, and then closes all the latter. I still believe that education should be able to teach people that greed for short-term gain is just as threatening to the common good as murder. I think education can show how this greed manifests itself in economic policies that encourage overextension, destructive farming practices, and ultimately the bankruptcy of not only our rural communities but all communities. As I write this, at the beginning of 2000, statisticians are reporting that for the first time in our history, personal savings have dropped below zero. We are literally living on borrowed time and borrowed money. Our inability to curb this kind of extreme greed is surely a failure of the kind of intelligence that education should foster.

These essays were written and first published over a period of twenty years, from 1980 to 2000. They do not follow here the order in which I wrote them. Of the earlier observations I made in 1980 or 1985 or 1990, few sound dated, as I feared they might. In fact, there is an uncanny similarity between my descriptions of farm economics in the early eighties and at the beginning of the new century. The age-old truth is once more confirmed: In a remarkable way history does repeat itself. How could it be otherwise when human nature continually repeats itself? Eerily similar articulations of the pattern of rural decline can be read in the history of England's industrialization in the nineteenth century, or in the decline of the Roman empire. If one wishes further proof, archaeologists find the same story in ancient Crete, Babylon, Cambodia, and Mexico. In all cases, the decline of rural culture ushered in a general decline. Are we genetically programmed to keep on repeating the same stupid mistakes?

Yes.

But my hope, my intention, is that this book exudes less anger or blame for the condition of rural society than it inspires love for the land and the human values the land nourishes. If that love prevails, it will lead more people back to the necessary virtues of survival fostered by rural life. Such a love could lead us to what democracy desperately needs to survive: a solid middle-class society of independent smallholders, not only of farms but of all kinds of businesses.

I also wrote this book to protest and reject the cultural images of farmers and rural people that public opinion continues to promulgate, especially through advertising and its influence on the media. The joys of rural life, though they bear little resemblance to the overblown fantasies of Rousseauist romantics, are still very real, and very much realizable, an antidote to the restlessness and chaos that infect modern life. But the vast majority of people don't know this. Rural life has been a victim of terrifically inaccurate media-imaging in our cultural history—as bad as the imaging of the Native American. Too often, farms are presented to an urban audience, already prejudiced against "local yokels," as places of discontent, boredom, poverty, crudity, despair, meanness, and ignorance. Surely, these tendencies exist in the cross-section of any community, including downtown Manhattan. But, as I know from hundreds of letters and conversations, just as there have always been people in farming who were unhappy and incompetent because they were not fit for it, so there are thousands, perhaps millions, of people in urban situations who are unhappy because they belong out on the land and do not know it. They have the true farmer's spirit in them, that blend of creative artistry, intellectual curiosity, independence, manual skill, and love of nurturing that marks the true farmer and gardener. If these people had been exposed to intelligent and craftsmanlike farming—instead of merely reading sour, Hamlin Garland–type "real life" novels about farmers who would have been failures at anything they tried—they might be living on and working their own little farms today. And with these hundreds of thousands of carefully kept homesteads gracing the landscape, the mad rush toward short-term money gain might be slowed.

I hope the readers of these essays share the vision I had as I wrote them: that producing some of our own food should be as much a part of life as cooking food ourselves. There is practical hope in that vision. Garden farms are even blossoming in cities today. In the heart of Cleveland, Ohio, in the shadow of skyscrapers, a team of horses patiently plow garden plots, and the National Guard, instead of cluttering up the highways on weekends with their "maneuvers," is hauling manure from the zoo to these new examples of urban farming. And it works to everyone's short-term advantage, too. This is what people can do when they put their wits together.

Sustainable farms are to today's headlong rush toward global destruction what the monasteries were to the Dark Ages: places to preserve human skills and crafts until some semblance of common sense and common purpose returns to the public mind.

Living at Nature's Pace

1

Green Fields, Red Ink

1986

As a boy some forty years ago, I could walk to Upper Sandusky, five miles from our farm, and never leave the land of my maternal kinfolks except for one small stretch. Rall family land. Good only for raising Ralls and Canadian thistles, people once said—a statement Uncle Carl loved to repeat bemusedly after his farm became the envy of any farmer with an eye to profit. There were sixteen Rall farms, if I remember, fourteen of them contiguous, averaging about 120 acres each in size. They had been divided up out of the acreage amassed by Great-grandfather and his four sons in an earlier era of big farming fever. I could walk, hunt, swim, fish, or play over that whole domain with my cousins and never fear a NO TRESPASSING sign, something I took for granted until many years later in Philadelphia, where I learned only the very richest could afford such a luxury. On those farms, about seventy-five of us, the fourth generation, were raised. We, our parents, and grandparents formed the society I knew. We played, worked, schooled, and churched together, and though there were ceaseless petty jealousies among us and a certain narrow-minded disdain for people who lived differently than we did—or even farmed differently—it was a securer life than most of us can hand our grandchildren today. And in the sense that ignorance can truly be bliss, a happier one. The self-reliant traditions by which the Ralls lived and farmed were more closely akin to the nineteenth century than to the twentieth. The Depression meant little to most of them—the farms either paid for or the notes carried comfortably by "Pop." Everything else we needed, the farms provided. And then there was always softball and Uncle Carl's fiddle.

The striking difference between that time and now is that today on the farm we anticipate change, are even surprised sometimes when it is not as cataclysmic as we feared it would be. But back then, no one suspected that in the twenty years between 1930 and 1950 the old Rall way of life would undergo more change than it had in the preceding one hundred years. Pistons would replace horses; electronics would replace thought. That earlier innocence brought a tranquility to our lives, turned our focus inward. In the assurance of stability, false as it was, the Ralls generated an amazing cornucopia on their little farms.

What I remember is farmsteads absolutely burgeoning with the thrum and cadence of life. Not only a passel of children underfoot, but a barnyard crammed with cows, horses, pigs, sheep, chickens, ducks, geese, cats, dogs, bees. From the woodland and the fields came meadowlark symphonies, bluebird poetry, owl nocturnes, and the thrills of the hunt—pheasant, rabbit, squirrel, raccoon, opossum, skunk, groundhog. There were orchards groaning with fruit, truck patches laden with potatoes and strawberries, kitchen gardens overflowing with vegetables, smokehouses full of meat, fish in the creek and the horse tank, pigeons in the hayloft. Every nook and cranny of the farm pulsated with life. Not only did this life press us with an eternal round of chores, but it also provided a kind of circus atmosphere of entertainment. On top of it all, peculiar perhaps to Rall farms (the other characteristics were part of most farms then) was a propensity for singing. Above the cackles, mooings, and squeals of the barnyard, a visitor might be startled to hear a bellow of human song, or a more feminine lilt from the kitchen. I often wonder now, if our lives were as drab and endlessly toilsome as the sociologists would have a modern world believe of farm life, why were my parents always singing?

The singing began to stop when the hucksters followed the plush prosperity after World War II to the farm, selling farmers a life like their "urban counterparts" supposedly led. (A farmer no more has an urban counterpart than a doctor has a crane operator counterpart.) As farmers expanded their acreage in the promise of living like urban counterparts, their young people left the land to *become* urban counterparts. Great-grandfather Rall begot four sons who farmed. Of their offspring, sixteen took to the land. These sixteen begot seventy-five children, but by 1975 only eight were farming Rall land. Three of the farms were sold out of the family, a travesty of the old Rall philosophy. One farmstead was abandoned, and several rented out. When the livestock were sold, the barns stood empty or were used to store expensive machinery needed scarcely a month out of the year. Groves and orchards were cut down to make way for more cash grain, the plow licking ever closer each year to the dooryard in a piteous, vain effort to wring a few more dollars' worth of grain from the land. Who needed orchards and gardens, anyway, when Kroger was there, already serving the urban counterparts? And then Uncle Carl, like my mother a few years earlier, fell out of his empty haymow, broke his neck, and died.

By 1975 the transition seemed complete. My father walked an empty, desolate barnyard, listening for the long-ago songs of life. He heard only a loose sheet of tin roofing, curled over, scratching itself distractedly in the wind. He cried. He cried because he no longer had the energy to keep the barn full of life himself. He cried because none of his children were willing to fill it full of life again. He cried because he could not die here on the farm amidst life, as his forebears had been able to do, but might soon, too soon, have to shuffle off to the county home like his urban counterparts.

Since most of the ballplayers are farmers, it does not take long for the after-game conversation at the Pour House Restaurant in Upper Sandusky to get around to the price of corn. It would not have taken long, anyhow, because many of the other customers are also farmers. In a nation where only 2.2 percent of the population farm commercially, Upper Sandusky is one village where at any public gathering as many as half the people will be farmers or the offspring of farmers. In 1985 farmer conversations follow a set pattern. For openers, someone trots out a new combination of vulgarities to describe the government's latest attempt to legislate a farm policy. Next comes a joke about a banker or FHA (Farmers Home Administration) loan officer, equally unprintable. Then the talk gets down to serious business. Who is going under next? Which bank is taking the worst beating? In calculating the number of farm bankruptcies imminent, a farmer will follow what might be called the "Rule of Distance." The farther away an area under discussion happens to be located, the more farm failures will be predicted there. Thus the farmers in the Pour House insist they can "hang in there with the best," even though all are small family farmers, the very ones the clergy and the newspapers lament. Each of them knows "one or two" of their neighbors who will go under, or who deserve to. In the county, they guess maybe "a dozen or so" will quit before the spring. But in Seneca County to the north, rumor claims that about 400 will soon be foreclosed on, and Crawford County is in even worse shape. By the time this kind of assessment has turned its attention to faraway Iowa, the picture painted is of a vast wasteland dotted with rusting John Deeres and shining Coors cans. After an appropriate silence to ponder that scene, a silence not altogether sorrowful, since it is difficult for an Ohio farmer to forget that Iowa's loss might be his gain, part three of the conversation pattern begins. This is the part usually accompanied by shouting and fist-banging. What is the reason for the farm dilemma? Ah, yes, the Cause.

"It was that damn grain embargo that ruined our foreign markets."

"Oh, hell, that's been disproved a hundred times and you know it."

"Well, you hafta admit it's the government meddling in our affairs that's the trouble. Get the government out, I say."

"You can't really believe that. We'd all be on the way to hell in a handbasket without subsidies. I didn't notice many of the free-enterprise Farm Bureau types turning down their fat PIK payments in '83."

(PIK, Payment-in-Kind, was a federal program that issued

certificates to farmers in lieu of cash as part of their subsidy on corn and wheat. The theory was that since the government had a huge surplus of corn, PIK was a way to get rid of some of it without paying out so much cash as in previous years. The program had no long-term effect in decreasing subsidies. Farmers and elevator companies would buy up certificates—you could redeem them for cash at the market price of grain at any time—in anticipation of a price rise. A lively trade in certificates sprang up and the complexities that followed are almost beyond comprehension.)

"By God, I didn't take no PIK-the-taxpayer's-pocket money. The rich bastards that did spent it on new pickups and satellite dishes."

"The real problem, boys, is interest rates. In the seventies, when the prime rate was eight percent, the inflation rate was over seven percent, so the real interest rate was less than one percent. Now the prime rate is twelve percent and the inflation rate is four percent, so we're paying over eight percent in *real* interest. I read that in some damn magazine."

"What do they mean when they say the problem is a strong dollar? The last time I had a strong dollar in my pocket, beans were eleven dollars a bushel instead of four, and by God, that wasn't no problem."

An older man at the bar has been listening to the flow of conversation with increased agitation. Finally he swivels on his stool and wags a finger. "I'll tell you what the problem is. It's us farmers, our own damn selves. When I was trying to organize farmers to control production through the NFO [National Farmers Organization], I learned. A farmer won't give up an inch of freedom for the good of the group. So let him hang separately. And then I learned something else that really scared me. You get a bunch of farmers who *would* hang together, and they would get greedy. They'd squeeze the consumer till he'd have to pay for food like he pays for cars and houses. Through the damn nose."

"Farming started to go downhill when they put lights on tractors," someone pipes up, trying to relieve the situation with a bit of half-humor. "Then we could go out and work all night like a bunch of fools and produce twice as much as we needed."

"My thinking is, if they would melt down every tractor that could pull more than a three-bottom plow, you would get rid of the groundhogs [large cash-grain farmers] *and* the surplus."

Surplus. There is that hated word landing right out in the middle of the table where everyone can hear it. It is easier for a nun to scream "dammit" in church than for a farmer to whisper "overproduction" in the bedroom. He knows that surplus is his immediate problem, but suggesting he cut production is like telling a ballplayer that over the fence is out.

"Cheer up. *Successful Farming* magazine says there's going to be a severe worldwide drought next year. *That'll* solve the problem."

Laughter. But not much. Everyone waits for someone else to comment. Finally, a woman who has said nothing previously speaks out. "Are you guys telling me things are so impossible that you would really welcome a bad drought?"

Silence again. Dave, the wryest of them all, stirs behind his can of Mountain Dew. "I know it sounds crazy, and it probably is, but no farmer is going to pull his own plug; and his banker wouldn't let him if he tried. The government's not going to pull the plug, because, among other things, it loves cheap food that surpluses make possible. So only nature can solve the problem and get away with it. Nature doesn't run for public office."

* * *

The current surplus—particularly in grain, but in other farm commodities, too—is awesome to behold. The corn crop in 1985 actually broke the old record of 8.2 billion bushels. Since billions hardly strike our shell-shocked brains any more than millions, contemplate for a moment how much 8.2 billion bushels is in bulk. Put that much corn in a bin having a diameter of thirty feet, and the bin would jut 2,747 miles into the air. Put that corn in cubic-foot boxes, and they would stretch to the moon and back eight times. The record-breaking crop is about 1.7 billion bushels more than the market requires, and added to last year's carryover gives us 3 billion bushels in storage. This glut sent prices skidding below two dollars a bushel. They rebounded to around $2.37, but since 71 percent of the 1985 crop is in the government loan program at the loan rate of $2.55 a bushel, the taxpayer is going to pick up another horrendous tab. USDA officials estimate that the subsidy outpay for crops alone (not counting dairy and other subsidies) will exceed the $18.9 billion spent in 1983, mostly on the notorious PIK program, which did nothing much to curb surplus production, either.

The wheat picture is even bleaker if one looks at it in terms of American agribusiness profits rather than relieving world hunger. While American yields were down a trifle, world production soared to a new record, led by China, once thought to be a potential big buyer of our surpluses. Russia also grew a good wheat crop. Europe, like Canada, has become a major exporter. Argentina is selling wheat at a price too low for American farmers to compete. India is now self-sufficient in wheat, and Pakistan nearly so. The countries that do need wheat are poor, and they will buy it, if at all, from the cheapest source, probably not the United States.

In the face of this "disaster," as the Farm Bureau is calling it, *The Ohio Farmer* magazine recently praised Ohio farmers for being number one in the corn belt in corn and wheat yields per acre this year—as well as tying with Illinois for the honor in soybeans. The Big Ten all over again. Perhaps a picture is in order: farmers dancing gleefully in the corn stubble with upraised forefingers, like victorious football players, while the sheriff hands them their foreclosure notices. "While it may not be in the best interests of agriculture as a whole," the magazine editorializes, "producing more is the goal of each individual farmer." To justify this almost unbelievable statement of agricultural suicide, the editorial then quotes an economist who also appears to believe that farming is only a matter of survival of the fittest, to hell with the soil, the money, and the society wasted in the process.

Such rugged individualism might have had some logic in the past, when there were still a considerable number of farmers too lazy (or too wise) to push for the highest possible yield—people who would rather have woodland to walk in and some good hunting cover on the back forty than win the county award for top corn yield. With such farmers around, the aggressive type-A's could overproduce without causing a nationwide glut. But the easygoing, "inefficient" farmers are mostly all gone now. Every acre a tractor won't fall off of is operated by a gung-ho farmer-tiger with a banker on his tail, in front of whom he must make his yearly Oath of Obeisance: "I do hereby solemnly promise to do everything in my power to raise more corn and soybeans than I did last year, no matter how much it costs me, so that I might enrich my bankers, my agribusiness suppliers, my government advisors, my farm magazines, and hopefully myself, while I claw my neighbors' eyes out." Priding himself on his "businesslike" attitude, he then goes forth to do what no businessperson would dare contemplate: to produce for a market, heedless of what that market needs.

This modern farmer is the best argument against the now popular notion that the farm problem is caused by "too many farmers." Wheeler McMillen, the celebrated Ohioan from Ada and dean of Ohio farm writers and editors, wrote a book in 1929 titled *Too Many Farmers.* He thought, during that financial depression, that getting the farm population down to 10 percent of the population would solve the surplus problems and leave enough dollars for the remaining farmers to share profitably. With the farm population at 2.2 percent of the whole today, I asked him what he thought now. "Now I'd write that there are too many acres being farmed," he replied. Precisely. That was true in 1929, too, and the nation would be further ahead right now if much of the highly erosive land had been put into forests in the thirties and kept there. If agricultural history shows anything, it is that as the number of farmers has decreased, production has increased. The reason is obvious: farmers who survive the suicidal competition are either more efficient managers, more ruthless miners of soil fertility, or better able financially to use new and costly scientific methods to increase production. To finance their expansion and their overhead, they must continue to apply the highest high-production techniques available, or themselves be taken over by someone who will.

Though taking land out of production sounds logical (and some will be removed by a paid conservation reserve in the new farm bill), removing a really significant amount of acreage is a highly unpopular idea in Washington. In the first place, it is doubtful that a deficit-ridden U.S. Treasury could afford the subsidy such removal would cost. More significantly, when you remove enough acreage to lower surpluses meaningfully, you don't just cut into farm income, but into the jobs and services of 23 million agribusiness workers, agriculture's support system. Better to lose a few farmers than to initiate an industrywide depression. Or so the thinking goes.

Nor is the farmer willing to cut production very much on his own. If debts do not lock him into overproduction, cultural habit does. But more important, the biological nature of his markets impels him to it. Nature is the fickle lady he must court. In 1983, a farmer bankrupt on 3,400 acres near Cridersville told me that he was making a personal study of other farmers in his dilemma. He found that the crucial straw that broke the farmer's back was not high interest rates, not low prices, not poor management, not high costs, not an overextension of credit—although all or some of these factors contributed to the downfall—but a streak of disastrously bad weather. If that uncontrollable risk is not bad enough, the farmer's destiny rides also on the whim and caprice of national and international politics. Be it weather or foreign policy, he must make his decisions six months to a year before he goes to market. The only way to face a future where the odds are stacked that much against him is with a full hand.

A farmer can only move at nature's pace. Although he is trying mightily, he can't make corn grow as fast as manipulated interest rates can make money grow. Corn grows the same whether interest rates are 8 or 15 percent. Ingenious man has conjured a way to make money grow as if it were alive—only money grows exponentially faster with age—but live creatures, biological lives, grow old and die. Not since the philosopher John Ruskin, who died in 1900, has society seriously concerned itself with the possible consequences of this discrepancy and asked basic questions: What rate of interest is compatible with biological growth? What rate of interest is compatible with a permanent, regenerative agriculture? And since all industry eventually rests upon biologically renewed raw materials, what rate of interest is compatible with a stable economic system?

The market glut and interest rates are not the only problems in farming, of course. The slowing of inflation, necessary to the health of the whole economy, cut the legs out from under those farmers and bankers who were trading on what they assumed was a perpetual increase in land prices. A classic boom psychology prevailed in the seventies. To get rich one merely had to borrow money, buy a farm, and when the price rose, use the paper gain as "equity" to buy another farm. Tax breaks encouraged the gambit. For example, in the early seventies, when prices were good, farmers were sometimes surprised to find at the end of the year that they owed thousands of dollars to Uncle Sam, a new experience. Their accountants told them there was an alternative. To avoid the tax, use the money as a down payment on another farm. Capital gains.

At the height of the boom, a farm in my county sold for nearly $4,000 an acre. A successful businessman in town who thought he might bid on the land, turned to me, calculator in hand, shook his head, and said, "There is *no way* that land can pay out at that price." But the voices of the few went unheeded. We do not teach the patterns of history in school nearly as well as we teach the patterns of football plays. The following quotation, for example, would strike most observers as a fairly accurate summary of what agriculture has been going through:

> Resumption of importations hurt U.S. manufacturing, but the downtrend was checked by European demand for American farm staples. As farm commodities soared and bank credit was liberalized, a wave of land speculation occurred. However, with the collapse of foreign markets, prices broke. . . . Land values sank rapidly, and numerous banks . . . failed.

Yet that description, from Morris's *Encyclopedia of American History,* describes the period 1815–21. Similar lurches between boom and bust occurred in 1837, '73, '93 and, of course, 1929—far enough apart to dim the succeeding generation's memories of the inexorable law of economics: There is no free lunch.

Another gathering of farmers. This one, not so coincidentally, also after a softball game. But the setting is entirely different. The game has been played on Abe Troyer's horse pasture in Holmes County and frequently interrupted by his Belgian mares tromping across the makeshift diamond. One of the teams is Amish, the other "English," as the rest of us are called in Amish country. As in the Pour House, it does not take long for the after-game talk to get around to the price of corn, etc., etc. But the Amish have a far more optimistic attitude.

"Crisis? The last three years have been nearly as good as any, financially, for us. Oh, sometimes an unwise Amishman gets in trouble, but we all help him out."

"But the university says you can't make a living with twenty cows. You need fifty to seventy at least."

"Oh, my goodness. If I milked seventy I'd be a millionaire."

"How much can you make?"

"We figure about half the gross income is profit, or maybe a little more. Twenty good cows should gross fifty thousand dollars today, and expenses, counting interest and taxes, are around twenty-two thousand."

"We English can't approach that percentage of profit. You must not be counting everything."

"Well, I bought a team of horses when I started farming in 1956 for eight hundred dollars. That team produced several more teams, and then I sold one of the original for four hundred dollars and the other for four fifty. You can't do that with tractors. My machinery is mostly old. The drill I use, my father bought in 1937. A few years ago I put seventy dollars worth of repairs in it, and I expect it to last the rest of my life."

Out of curiosity I gave one Amish farmer the Ohio Extension Service (OES) expense budget for corn for 1984 and asked him to compare his costs with those calculated by the economists for a typical yield of 150 bushels per acre, which the Amishman said his crop made in 1984. The comparison looked like this:

Item	**Typical Cash Grain**	
	FARMER'S COST *Per Acre*	AMISH COST *Per Acre*
VARIABLE COSTS:		
Seed	$ 24.00	$18.66
Purchased fertilizer	$ 63.00	$ 9.10
Lime	$ 8.00	$ 5.06
Pesticides/ herbicides	$ 28.00 [1]	$ 2.50
Fuel, oil, grease	$ 19.00	$ 3.00 *(est.)*
Corn drying (fuel and electric)	$ 23.00	$ 0.00
Trucking fuel	$ 3.00	$ 0.00
Repairs	$ 13.00	$ 0.25 *(est.)*
Misc. supplies, utilities, soil tests, small tools, crop insurance, etc.	$ 13.00	$ 0.50 *(est.)*

Item	Typical Cash Grain	
	FARMER'S COST *Per Acre*	AMISH COST *Per Acre*
Interest on operating capital	$ 12.00	$ 0.00
FIXED COSTS:		
Labor	$ 9.00	$ 0.00[2]
Machinery charge	$ 50.00	$ 5.00[3] *(est.)*
Land rental charge	$110.00	$ 0.00[4]
Management charge	$ 18.00	$ 0.00[5]
TOTALS:	$393.00	$44.07

1. Herbicide costs could be as high as $70 per acre.

2. The Amish farmer hires no labor and says his own labor is part of his profit, not part of his costs.

3. The Amish farmer does not know how to figure this because most of his machinery is so old "it is actually gaining in value now," he says.

4. See next paragraph.

5. "What is this?" the Amishman asked slyly. "The time spent asking experts how to farm?" Again he figures this as part of his salary, not as a cost.

With the price of corn reckoned at $2.40 a bushel for 1984, the typical farmer on the OES budget grossed $360 per acre on a fine crop but ended up losing $33 per acre. In Amish accounting, the same yield brought a net profit of $215 per acre. Even if you allow a portion of that to go as fixed cost for owning the land, as in English accounting, the difference is still well over $100 per acre. What is just as important, the Amish seldom sell corn, but feed it to livestock and sell it as milk, meat, cheese, and eggs, thus retaining an even greater share of the corn profit dollar.

I told the Amishman that he needed to add a cost for cultivating weeds in the corn, in lieu of more herbicides. He thought that might add another dollar or two per acre using the horses and his ancient cultivator. And what about the cost for spreading manure?

"Yah," he replied, nodding in a way that revealed he had thought a long time on that matter. "Farming is too complicated to arrange neatly in figures. When I'm hauling manure, should I charge that to cleaning out the barn, which keeps the cows healthy, or to fertilizing the field, which reduces the fertilizer bill? And the manure does more than fertilize. It adds organic matter to the soil, which helps it to use soil nutrients better and soak up rain better to reduce erosion. How much do you charge for that in your computer? Or maybe I should charge manure hauling to training the young colt pulling the spreader, or to winter exercise for the older horses. Or maybe I should deduct manure from machinery wear because the ground gets mellower with manure. I don't know how to calculate that accurately. And if I'm hauling manure, I'm not up at the sale barn wasting time or buying something I don't need."

Many conservative "English" farmers nod assent to the Amish low-cost farming methods and insist money can be made in farming even in the eighties. A more accurate statement of the problem, they say, is that one may not be able to make money farming the way one wants to, which is another story. One farmer, who refuses to be quoted by name, says: "We have a whole generation of farmers raised in comparative luxury who think farming means driving big tractors around five months of the year and bragging about it the other seven." Survival today may mean a lower standard of living, a hard pill for farmers to swallow after thirty years of being told they could live like their urban counterparts. But the Amish farmers who are doing well can hardly be said to live poorly. Not the ones I have visited. They have plain but comfortable, substantial homes. Gas refrigerators, water heaters, and stoves. Diesel motors run generators for the milking machines and milk cooling tanks. They even use tractors for stationary power to run grain grinders and silage choppers, though not for field work. Field harvesters are pulled by horses but run by gasoline engines. The Amish will use technology to lighten their work but not in a way that would tempt them to expand farm size on a grand scale, which is how the trouble got started in the rest of agriculture.

Dave McCoy, near Fredericktown, is not Amish, though he admires their farming. (His secret desire is to own an airplane.) At thirty-six, having farmed the home place for seven years, he knows the odds are against him as a small farmer, but he accepts patched clothes, an old truck, machinery he often makes himself (one of his tractors has parts from a dozen others on it), and long hours. That's the price he must pay to be in business for himself with his thirty-six-cow dairy operation and to provide a good home for his wife, Marcia, and their three small children.

McCoy is the small commercial farmer's most articulate champion in Ohio. Don't ask him by long distance how to solve the farm problem, because it takes him an hour and a half to answer. He loves to make controversial statements, then flash his impish, disarming grin. Example: "The price of milk *should* be low. That's how economics works. There's too much milk. Subsidies only worsen the problem." Impish grin. "And it wouldn't be the little dairy farmers who would fail if subsidies were removed. It would be the big boys. That's why subsidies won't be removed." Impish grin.

As an honors graduate in agricultural economics at Ohio State, McCoy drove his professors to despair by coming back each fall from summer jobs in agribusiness, pounding his fist on their desks, and protesting: "You aren't teaching me what I need to know."

"I might not have come back into farming at all," he says now, "if it were not for working with some black small farmers in Mississippi after I got out of the Air Force. I went down there in a church-related program. Contrary to what I'd been taught about such farms, those guys were making money. They were smarter than the professors."

Trying to farm small was difficult enough, but McCoy made it even harder on himself. He decided to farm organically—no purchased fertilizer or pesticides at all. "The Spray brothers at Mount Vernon are very successful organic farmers, so I thought, Why not try it? I wanted to buy the farm from my father, but fortunately he ignored me. Had I done that, I'd be washed up now. Instead, we worked out a rental arrangement we could both afford."

Maintaining soil fertility with manure and legumes instead of expensive fertilizers did not prove difficult, but only after six years has McCoy found an effective way to control weeds in corn by cultivation only. Now his production costs are almost as low as those of the Amish. By using old and rebuilt machinery, he says, he has only $6,000 invested in tillage equipment. "If all goes according to plan, we will be out of debt in less than two years," says Marcia.

McCoy's experiences in accounting, farming, and working in agribusiness, has convinced him that small farms, not large, are the way out of the dilemma. "I think maybe the economies of scale work backwards in farming," he says. "Big farms can't be efficient. I always think of a big dairyman I knew who made the cover of a farm magazine standing in front of his expensive one-hundred-cow facilities. What the photographer didn't know was that behind that fancy, big setup there were dead cows lying in the manure." He ticks off his cures for a sick agriculture. "Get more people on the land. Pride of ownership will solve erosion problems better than an army of soil conservationists. Absentee ownership is a grave problem. I understand forty-five percent of Ohio farmland is absentee-owned. Secondly, diversify production, and bring back local marketing. Put livestock and legumes on *all* the land. And treat animals the way animal rightists say we should." Impish grin. "It will make us more money in the end. Thirdly, get rid of the absurd notion that farmers should live the same lifestyle as urban people. Fourthly, don't buy anything you can produce for yourself. And fifth, quit listening to so-called experts. There's not half of the economists at Ohio State who can run a good cash-flow analysis." Impish grin.

Truth floats phantom-like, just beyond the grasp of fact and statistic. Urban society, as Marcia McCoy observes of her acquaintances in Boston, Massachusetts, her hometown, have little inkling of what farmers are going through. Nor will they become concerned until they realize that this is no agricultural crisis but a money crisis. Bankers are sweating more blood than farmers. By the end of 1985, 437 farm banks (banks that have more than 25 percent of their loans with farmers) were on the FDIC's list of "problem banks"—up from 231 a year ago. Sixty-two of the 120 banks that failed in 1985 were farm banks. Ohio, with a balanced farm and manufacturing industry, is in better shape than most midwestern states, but Iowa's superintendent of banks, Thomas Huston, was quoted in the *Wall Street Journal* as saying, "We're in for a hell of a whipping." If Iowa gets whipped, can the rest of the country be far behind?

Yet 42 percent of Ohio's farmers have no debt at all. Many of them are ready to buy farmland, and they wait only for the price to fall still lower. National surveys indicate that only about 9 percent of farmers are in drastic trouble, but farmer surveys tend to emphasize the positive. In a survey in *The Ohio Farmer,* 93 percent of the respondents said they planned to stay in farming, yet 49.5 percent described themselves as "Not good but making it," which to me means they are in trouble. Most farmers I talk to believe "the big boys are in the driver's seat," but statistics do not bear this out. Economist Allan Lines at Ohio State points out that a debt-to-asset ratio of over 25 percent raises a small red flag, and one over 50 percent raises a large red flag. The two groups of farmers with the greatest number of individuals in the over-50-percent category are either younger than thirty-five or have over $200,000 in gross annual sales. John Marten, the respected columnist at *Farm Journal* magazine, wrote last summer: "It is folklore to say big farmers are doing well. Those grossing over $250,000 made up less than twelve percent of our survey, but they owed thirty-six percent of the debt."

All economists seem to agree that smaller farmers with off-farm income are the most secure financially. But since they are in many cases only financing their farming habit with off-farm income, their farming can't honestly be described as a healthy business.

A man who describes himself as a consultant in farm-business planning stops by to visit. Once he was an Extension Service advisor to farmers, then decided to take a crack at his own advice and try farming himself. Becoming disheartened last fall, he sold out. Now he teaches classes in how to survive on the farm—he seems to see no irony in that. "You can't quote me," he says, leading off with that now familiar refrain, "but beware of statistics put out by agribusiness. Mental attitudes are so negative right now that business leaders fear any bad news will send land prices crashing and banks and insurance companies will lose billions." Farmers owe bankers and others $198.6 billion dollars. The Food and Agriculture Policy Research Institute at the universities of Iowa and Missouri foresees loan losses as high as $20 to $25 billion by 1993 if the present economic conditions hold. The bumblebees and the McCoys continue to fly.

Walking Rall land today, I can find an exception and a proof for every statistic being bandied about. Asked by people from other regions if the farm problem has reached crisis level here, I open my mouth and nothing comes out. Generalities fail me. It depends on which particular farmer you want to discuss. Each situation is a little different. Of the eight fourth-generation Ralls farming in 1975, three would be considered large farmers—over a thousand acres in cultivation. One of them has failed; one is worried; and the third waxes financially stronger. A fourth farm is a midsized, father-son arrangement, the very kind supposedly in greatest danger. But this farm is not in financial difficulty. The other four never expanded much, were never the kind to attract a farm editor's attention. But they survive, following a lifestyle more reminiscent of the old Ralls than the new. One of them has never bought an acre of land but farmed the home place, taking over from his father when he died, dividing the yearly profits with his brothers and sisters, saving his pennies painstakingly over the years. He is the securest of them all today. I have a hunch that these eight farmers provide about as accurate a cross section of Ohio farming today as any.

But that is not the end of the story. Something else is stirring in Ohio's rural areas that is not so gloomy. While tears (mostly of the alligator variety) are being shed for the passing of the family farm, a new kind of rural household is rising in its ashes, one that retains most of its benefits and few of its disadvantages. From a peak of 6.81 million in 1935, the total number of farms has fallen to 2.43 million—43,770 were lost last year, and 147,880 since 1980. But according to the Department of Commerce, the number of farms of under fifty acres has been *increasing*—by 17 percent annually between 1978 and 1983. That figure does not include the myriad small homesteads that have little or no sales. Nor are the 33 million gardeners counted, even though some imaginative statistician has noted that the dollar value of their production in aggregate was greater last year than the net income from all commercial farms. If we were to change the definition of "family farm" to include backyard garden-rabbit-chicken-workshop operations in town, one- to five-acre suburban minifarms, fifty-acre hobby farms in the country, and hundred-acre-plus, part-time commercial farms financed by off-farm work, then this cherished institution is not dying but growing by leaps and bounds. These households foster, as much as the old family farm did, diligence, personal responsibility, love of nature, social stability, and the other traditional rural virtues.

I am suspicious of statistics, but in the case of small farms, they reflect what is now happening on old Rall land. *Since* 1975, nine more fourth- and fifth-generation Rall families have come back to tend ancestral soils, bringing the total to nearly what it was in 1945. One newcomer is a full-time farmer having land elsewhere in addition to Rall holdings. Four others operate small commercial farms but have other jobs or businesses in addition. The other four tend from one to thirty acres each, hoping to make a little money at it but mostly because this kind of country life suits them. The newcomers have brought improvements to their holdings that would make the old Ralls smile. They have set out more orchards than were stripped away from the old farmsteads. They have saved woodland that was destined for the bulldozer. They have allowed new woodland to grow up. They raise huge gardens. Old pastures have been improved; new ones planted on land that should never have been cultivated. In old barns and new, livestock once again brings farm song to the land. There are ponds full of fish, hives full of bees, a smokehouse full of meat. And there are children always about, fishing, swimming, working, laying out baseball diamonds in cow pastures once again. A new rural community.

It is not the same, but the differences are mostly to the good, I think. The old isolation is gone, the old innocence. The lifestyle appears suburbanish, and is to some extent; but the talk at gatherings is invariably of country matters, and the moral stance inveterately on the side of conservative rural virtue. If a vigorous rural society is one of the strengths of a nation, it just may be about to flourish again. And it won't be the result of government subsidy, church project or educational program, but in spite of them.

The current love affair that economics is pursuing with large-scale farming is by no means its first, nor is there any indication that this one will be any more permanent than the others. As the population increases, farm size, over the long term of centuries, decreases. When Charles Rall and his sons were building toward their eventual 2,000 acres, his neighbor to the south, Dave Harpster, was building an empire of 7,000 acres. In the latter half of the nineteenth century, bonanza farms, as they came to be called, flourished in the western corn belt, in California, and particularly in the Dakotas. Eighty farms in the Red River Valley were over 3,000 acres in size, and some much larger. When the Dalrymple Farm reached a peak of 75,000 acres, it was employing 1,000 men, 800 horses, 200 mechanical reapers, and thirty steam-operated threshing machines. These farms were built on cheap labor, cheap land, and cheap taxes, just as today's were built on cheap energy, cheap money (from inflation), and cheap taxes (in the form of tax breaks).

The bonanza farms barely lasted into the first quarter of the twentieth century, when land, labor, and tax costs soared. In years of poor weather conditions, which were common in the Dakotas, they proved no more profitable than small farms anyway, and labor problems were almost as vexing as slavery in the South. Nor did the huge farms, then or now, foster a vigorous local town and country economy. When the Dalrymple bonanza sold out in 1917, William Dalrymple wrote in a letter to an acquaintance what might very well be one of the wisest lessons a civilization can learn:

> My brother and I have decided to give up operating the farm and divide it into small farms. . . . It is better that this be done for many reasons. . . . Big farms were once good for publicity. But economic conditions have changed. . . . It will be better for the state, for the towns and cities . . . to have a great many small farms in place of one big farm.*

We have been there before. Will we learn this time?

*Hiram Drache, *The Day of the Bonanza* (Fargo: North Dakota Institute for Regional Studies, 1964), p. 211.

2

For Amber Waves of Green

1980

Three farmers stand gawking up at the huge Steiger farm tractor at the Wyandot County Fair. Douglas Steiger invented the monster for the endless plains of the Dakotas and the Red River Valley, where you could fall asleep plowing and wake up two miles later without having hit anything higher off the ground than a fishworm. The articulated steering, four-wheel-drive monster weighs twenty-two tons. It is too wide to fit safely on any road in Ohio and it seems as out of place at an Ohio fair as an antiballistic missile. Four people could crawl inside the cavernous hubs of each set of ten-foot-tall dual wheels. One ladder leads up to a kind of foredeck alongside the 450-horsepower motor. A second ladder ascends to a commodious cabin outfitted with air conditioner, heater, stereo radio, tape deck, upholstered bucket seats, banks of blinking electronic lights that monitor every working part of the tractor and ranks of hydraulic pump levers that give the operator finger-touch control of enough strength to wrench a good-sized tree out of the ground as if it were a pigweed. A large sign hangs on the side of the tractor. It reads: "THIS VEHICLE COSTS $130,000."

The price tag is meant to impress city folk with the high cost of farming. City folk are supposed to believe, looking at the sign, that they get a real bargain on food. But Wyandot County villagers passing by seem unimpressed. The children have seen more wondrous machines in *Star Wars.* Their parents are divided into two camps: one believes no farmer can afford such a tractor; the other thinks all farmers are "rich sonsabitches." Both camps move on to ogle the snowmobiles.

Only the farmers stand in reverent awe. Two of the three beneath the brute at this moment can remember the feel of horse reins in their boyhood hands. They gaze up at the tractor and the sweet visions they have dreamed all their lives—of unbridled technology—pass before them. They do not think any farmer in Ohio will buy such a technical marvel, but it is satisfaction enough just to see it. Like Moses, they have looked at the promised land though they will never enter it. They cannot find words to speak. One shakes his head. The other kicks the tire that dwarfs him. The third, standing a little off from the others finds his voice: "How many people you reckon that thing'll sleep?" he grunts. He can afford to be a bit blasphemous before the graven image. He intends to buy it.

By coincidence, these three farmers represent the three economic groups into which Ohio farmers can more or less be divided—by coincidence, because the three groups rarely meet socially. The farmer who intends to buy the tractor farms 3,300 acres and does not want to be called "farmer" anymore. He is not too keen on "agribusinessman," either, but that will do. " 'Farmer' implies something that I'm really not," he confides earnestly. But he mentions that only where the other two won't hear him, as he is afraid such sentiments sound like "puttin' on airs." An Ohio farmer would rather cut off his fingers one at a time than be accused by other farmers of putting on airs. Besides, using the word *agribusinessman* will upset the one farmer standing next to him who farms only 120 acres. "I'm a farmer, a dirt farmer," Mr. 120 Acres likes to say. "I *farm,* not just to see how many acres I can race a tractor over every spring and fall. You can call 'em agribusinessmen if you want, but I call 'em groundhogs. I can make a living off the land they use just to turn their big rigs around on."

The third farmer, who farms 800 acres, laughs at that remark but not very wholeheartedly. He secretly wishes Mr. 3,300 and Mr. 120 would both quit farming and rent him their land so he could be a "groundhog" too and get with the corn, soybeans, and "Florida rotation" that Mr. 3,300 has perfected. "At least I'm a full-time farmer," he says proudly when neither of the other two is present. "I don't finance my farming with another business or another job like they do."

About the only characteristic the three farmers have in common is that they will not talk for publication. Or at least they will agree to be quoted only on unimportant questions. Everything else is off the record. They have picked that up from watching television newscasts. They have learned how to handle the press. For example, two of them are millionaires, but to admit that, to be quoted saying that, would be puttin' on airs. They are millionaires only off the record, where neither farmers nor the IRS is listening.

"Sure, we're millionaires," says Mr. 3,300. "On paper. With the price of farmland at $3,500 an acre around here, you're a millionaire even if you've got only 300 acres paid for. You can become a millionaire without even getting out of bed in the morning. But so what? A millionaire farmer these days doesn't have enough cash between planting and harvest to drive his Cadillac to town to buy an ice cream cone."

Mr. 3,300 belongs to a select group of farmers that the U.S. Department of Agriculture categorizes as selling "in excess of $200,000 in farm products annually." The USDA puts the number of farms in this category at over 63,000 nationwide, quadruple the number ten years ago. But as ballpark figures, both are well on the conservative side. There are enough farmers in Marion, Wyandot, Hardin, Hancock, and Seneca counties with gross sales in the half-million-dollar range to fill a Greyhound bus—and these counties are not even in Ohio's top ten counties for total cash receipts.

Not counting large greenhouse growers and other large specialized fruit, vegetable, or ornamental businesses, the "over $200,000 club" in Ohio is composed mostly of large grain farmers. There are a few huge livestock operations up there too, but most of the real high rollers deal in soybeans (now the top money-maker in Ohio agriculture), corn, and wheat. Livestock enterprises are difficult to expand to the grand scale that the biggest farmers like to operate at, mostly because of potential health hazards and costs from climatic conditions in Ohio (too cold or too wet much of the year) and difficulties with odor and water-pollution regulations. Besides, animals mean a farmer must stay home twelve months a year. Grain farming, on the other hand, is a six-to-eight-month operation, like football. Grain farming is clean work, and far, far easier than husbanding animals.

"Paradise for the groundhogs," says Mr. 120, "would be twelve rows of corn from here to Florida that they could plant coming north in the spring and harvest going south in the fall."

The successful large cash-grain producers are getting rich on paper because they have steadfastly adhered to a simple rule: inflation is a constant in the economy, and "the Lord ain't makin' no more land." Therefore, the rule goes, buy land when you can and don't worry about the price. Keep yourself as far in debt as possible without going under, and inflation will make you rich.

Accompanying this prime rule of thumb are a few enhancing practices. One: if you can't buy land, rent it. The more land you farm, the more you can spread costs, and renting gives you an inside track if the land comes up for sale. Two: be nice to everyone, especially retired landowners. You never know who you might need to rent land from in the future. Three: do what you can to gain vertical and horizontal control of your grain business. Buy a machinery dealership to get equipment at wholesale. Buy a grain elevator to gain more marketing leverage. Buy a bank, or at least a banker. Four: hire good men and pay them well and they will make you richer. Five: operate another business to keep help employed in the off-farm season.

"People around here think because I farm 3,300 acres that I will stop expanding," says Mr. 3,300. "I have only begun. I have no choice. You either keep expanding or someone else will, and you lose out. It has taken me this long to get a million dollars in debt, but I would not hesitate, if the right opportunity presented itself, to borrow another million tomorrow flat out. Your ordinary farmer doesn't understand what's happening. A hundred thousand farmers could handle all the land in the United States and probably will. I intend to be one of them, if I can. But a million dollars is nothing. A million goes through my hands every year now. I can hold on to damn little of it, but the equity mounts up. Becoming a millionaire in farming is easy. Never in history has a business gotten so much free help from its government and its scientists. The government, in trying to help small farms, has been the single most helpful partner to the big farmer. But it is no use talking about it. Some people can see what's happening and some can't."

For all that private bravado, big grain farmers like Mr. 3,300 keep a very low profile in their communities, where they are more often resented than admired. They socialize with urban businessmen with whom they have much more in common than with farmers, or with large landowners in other states or counties who are not in competition with them—yet. Blocked from buying more land in their own communities, Ohio high rollers are buying it, in thousand-acre chunks, where they can: Texas, North Carolina, Florida—wherever the land has potential the natives haven't yet understood, so that the price is right. But even at home, their empires tend to grow because smaller farmers can't outbid them at land sales. "Yeah, a lot of them hate me," says Mr. 3,300, "and I suppose you can't blame them. But it works both ways. When those small farmers who bad-mouth me retire, they rent their land to me because they want to go with a winner."

When Mr. 800 heard that Mr. 3,300 said it was easy to become a millionaire super-farmer, he laughed hollowly. "That son of a bitch," Mr. 800 said. "Ask him what will happen if inflation takes a nosedive. If we have a Depression, he'll be ruined overnight. You ask him where the hell he got the money to move as fast as he has. You'll find out that almost every one of those high flyers got a big chunk of money from somewhere else right when they needed it. Now, maybe they got it by being smarter than me but they by hell did not get it out of farming, because you cannot get rich fast from farming. He's playing the game of survival of the greediest. We all have been doing that. 'Get big or get out.' That's shit. We can't all get big, because there's only so much land. We thought if we squeezed out the poor farm managers and all of us ended up with 800 acres, everything would be hunky-dory. Well, 800 isn't enough. I need more land now. I'm *hurting.* So's everyone else. When land does come up for sale or rent, the well-heeled big boys can outbid me. But by God, not because they are better *farmers* than me. In fact, they are a lot harder on the land than I am. But they can sweet-talk the bankers better."

Mr. 800 is representative of the middle-income farmers. His bitterness is also representative, although that is the best-kept secret in farm journalism today. Agribusiness does not want to admit that the fiscal philosophy it has convinced farmers to follow, being predicated on unlimited expansion, is a farce—at least for farmers. Not all middle-income farmers are as bitter as Mr. 800, however. Many, perhaps most, have come to terms with the harsh realities of economic expansion, and, by diversifying grain production with varied livestock enterprises and long working hours, they make a moderately comfortable living. Others farm their acreages as straight grain enterprises but start other businesses or jobs on the side. The business may be as ordinary as trucking or as novel as a school of exotic dancing. They learn, as the super-farmers have learned, that the entrepreneurial skills that make them successful farmers can make them successful at almost any business.

The USDA defines the middle-income group as those farmers selling between $20,000 and $200,000 worth of farm goods a year. They account for 51 percent of all farm production and number about 837,000 nationwide. These are the farmers one is most apt to think of as "typical" farmers. There is not much about them that's typical, however, and within such a wide range—$20,000 to $200,000—few generalities are valid, other than that most of our meat, milk, eggs, and other animal products comes from this group. The larger operators tend to combine corn and hogs, or corn and poultry, or corn and cattle feeding. The smaller farms are often dairy farms: income on dairy operations is very high relative to the number of acres. Small Ohio hill farms may be sheep farms, beef herd farms, or in feeder pig production. It is interesting to note that the three Ohio counties leading in total cash receipts are Mercer, Darke, and Wayne—counties known for hog, cattle, dairy, and poultry farms, not huge cash-grain operations.

Farmers in the middle-income class are not as heavily indebted as in the top class, but the degree of debt among them varies widely. This fact accounts for the difference in opinion among farmers about how good their business is. Two farms identical in size and operation are worlds apart economically if one is heavily indebted and the other not. One farm may be paying out $25,000 a year in interest alone, $25,000 the other farm can count as profit. It was the big borrowers in the middle-income group—seduced into that position by the blandishments of agribusiness but unable to expand their land base like the super-farmers—who comprised the majority of the protesters who drove their tractors to Washington last winter (1979) to bring attention to poor farm prices.

"The government is so worried about helping the smaller family farmer, or pretends to be," says one farmer. "Well, I'm a smaller family farmer, and because I never listened to government and farm-magazine bullshit, I don't need any government help. My farm is paid for and my cows are good ones. I make all the money we need from 240 acres and sixty cows, thanks to the good

Lord. We don't live so high and mighty but we don't owe nobody any money. We will be around long after the big spenders go broke. We don't believe in getting big, just better."

There is discontent, alarming discontent, among middle-income farmers who have had to borrow heavily in today's economic climate to stay in farming. Their discontent, even in the face of tolerable profits and tremendous growth in equity, can shock a salaried urban worker who is building little or no equity other than in his home. One young farm couple, still in their twenties but showing financial growth double that of urban executives twice their age, describe their discontent this way:

"Our idea was to have a nice farm of about 500 acres where we could raise our kids, work hard, enjoy what we were doing. Well, we have the farm and we are getting it paid for little by little, but we can't relax. We can't stay at this plateau and just improve what we have. Inflation won't let us. If we don't expand, we will slip backwards. So how should we expand? We can raise more livestock. That means a big investment in buildings and probably hiring someone. The hog market doesn't look that good. On the other hand, if we buy more land at the going price of $3,000 to $4,000 per acre, we take on a terrible risk in our opinion, what with all this talk of depression. But the point is, we don't *want* to be great, big, fat farmers. We just want what we have. If we can't do that, we're thinking about selling out. If we did, at today's crazy land prices, we could pay all our debt and still have $300,000 left over. We could invest that, get easy jobs in town like other people have, nine-to-five, and no worries, and maybe have a lot better life than we do now. Why should I be out there in the barn freezing my ass in the middle of the night with them damn pigs, if I can't have a secure farm to go with it? If those pigs were raised with union labor, the consumer would pay seven dollars a pound for it. If the government wants to regulate the shit out of every move we make, who needs it? Let the government raise pigs."

Mr. 120 Acres belongs to the largest group of farmers—1.8 million in number according to the USDA—with sales of less than $20,000 annually. In many ways, farmers in this group seem to be the only happy ones, even though they put in more hours than either of the other groups. Mr. 120 works a factory shift and farms at night. His little farm is full of sheep and hogs. There are chickens in the coop, a few steers in the lot, and a barn full of hay. His farm looks the way a farm used to look. He farms by preference. He doesn't feel trapped. "I used to have a plan to get more land and go into full-time farming," he says. "But land prices always rose faster than I could accumulate money to buy with. I finally decided that I had the best way of life—working in town and farming, too. My farming takes up most of my so-called leisure time, but I get a lot of satisfaction out of it and a pretty good side income, too. I can make a little money because my expenses have not risen the way they have on bigger farms. Heck, I don't borrow money. I don't buy new equipment. My tools are the same old ones I used twenty years ago. I can fix them in my shop. The only real increase in cash layout for me is fuel. I've got manure and I rotate clovers—don't need to buy much fertilizer. I cultivate weeds instead of spraying them and save thirty dollars an acre on herbicides. It doesn't cost me much more to raise six-dollar-a-bushel soybeans now than it did years ago to raise three-dollar-a-bushel soybeans. I don't envy the big boys. I wouldn't have the guts to go in debt like that. I can't understand how anyone would want to live like that. It would kill me. The ones I know—I think it's killing them, too."

Farmers who take on a job in town make up the bigger portion of this third group, but there is another, more recent, type of part-time farmer much in evidence in Ohio. These are people with established careers in other professions or businesses who are buying small farms to live on and, they hope, also make a profit on. Somewhat surprisingly, they add an optimistic glow to the future of farming in Ohio. Bernard Erven, agricultural economist at Ohio State, has been studying part-time farmers and concludes that they are not poor, not failures, not necessarily even low-income, not expanding acreage, but definitely not leaving farming, either. Erven's survey shows that the part-timers are in a stronger financial position than many full-time farmers, and can more easily survive shocks from low prices or bad weather.

To better understand why the lower-income farmers are satisfied with their lot while the budding millionaires are not, one has only to look briefly at the economic situation in Ohio agriculture (which is reflective of the situation across the United States). Over the past twenty years or so, net farm income has barely doubled while costs have more than tripled. The tremendous surge in gross income in the seventies was eaten up by an equally tremendous surge in expenses. Since 1975, net income has actually been declining. The USDA says that costs went up another 16 percent nationwide in 1979, and the department is gloomily projecting at least a further 10 percent increase (some say as much as 20 percent) for 1980, without a concomitant increase in gross income. Agribusiness leaders are protesting that the USDA projections are unduly pessimistic, but farm advisors are telling farmers to cut back on machinery purchases and other supplies where at all possible.

The business picture of farming in Ohio may be worse than the statistics at first indicate. Production expenses in 1978 were 84 percent of gross farm income, the largest ever. But that still seems to leave 16 percent for profit, not bad for a larger corporation. But a comparison of production expenses with total cash receipts reveals that expenses in 1978 were *99 percent* of cash receipts. In other words, the only appreciable net income, statewide, was due to government payments, nonmoney income, and other farm income. (As Mother used to like to say: "There's a living in farming, but about the only profit you have left at the end of the year is the manure.") The only news that puts a rosy glow to these figures is that the paper value of land has been increasing even faster than the costs of farming. But the only way a farmer can get his hands on that paper value is to sell the land and then watch the government take out a big tax bite.

"I'll tell you something," says one farmer. "Everyone says you can't afford to buy this high-priced farmland. Well, then when you retire and need some extra cash, you can't afford to sell it, either."

The vulnerability of farmers and farming under the burden of current economic pressures is not recognized by most bankers yet, or at least few will admit it. After all, productivity in American agriculture is still very high, while productivity in other sectors of American industry has been declining. Without this productivity, inflation would be even worse and the United States would have little to keep the overseas balance of payments from going even more out of whack. (In 1978, Ohio was the tenth-ranked state in exports, with $1.1 billion worth of farm products. While there was a negative nonagricultural trade balance of $48.2 billion in the United States that year, agriculture had a *positive* trade balance of $13.4 billion, demonstrating how important Ohio farming, and farming in general, is to the nation.) But regardless of the ultimate effects of high production in agriculture, neither banker nor politician really wants to do anything to slow it down for fear economic chaos would quickly follow.

And if farmers are vulnerable economically, their soil has become even more vulnerable biologically. No ear of corn has ever heard of 6 percent interest, much less 12 percent. To keep pace with that kind of inflation, the farmer is forced to squeeze every kernel of production he can from the soil. He increases applications of toxic chemicals known to adversely affect soil microorganisms. He buys heavier and heavier machinery, which compacts his soil, so that yields drop. He applies increasing amounts of fertilizers to bring yields back up. These stratagems all seem to reach points of diminishing returns. The bugs become immune to the insecticides, and scientists are now reporting strains of weeds immune to herbicides. Adding more fertilizer to the soil no longer increases yields to the degree it once did.

Coming to grips with these problems is not easy for agricultural colleges or agribusiness suppliers who have stoutly maintained that American soil, climate, and ingenuity can continue to expand cheap food production indefinitely. But Sylvan Wittwer, assistant dean of the College of Agriculture and Natural Resources at Michigan State, writes in a position paper prepared for the Hudson Institute:

> A dramatic change is now occurring. Yields of all major U.S. crops (corn, wheat, sorghum, potatoes, cotton, sugar beets, field beans, peanuts, soybeans have reached a plateau, following a precipitous rise during the 1950s and 60s. Prices for farm products have become volatile, unstable and more uncertain than ever. Meanwhile, energy-intensive farm inputs have risen sharply with a continuing upward trend. Thus American ag-

> riculture, the farmer and the consumer are in a uniquely vulnerable position with no counterpart in history. Increased profitability for the farmer and reasonable prices for the consumer are both predicated upon sustained increased production.

Among the factors contributing to the leveling off of farm production, soil erosion is the most pervasive. While Ohio's average rate of soil loss is only slightly above the safe rate for all cropland, it is two or three times the safe rate for *sloping* land, says the federal Soil Conservation Service. That loss costs Ohioans an estimated $230 million annually. (It's worse in Iowa, if that's any consolation.) Yet farmers continue to clear hill land and farm it because expansion is the only way to increase income quickly. The so-called new conservation-tillage methods do not work well on cold clay soils in Ohio, either, and depend on heavy applications of toxic herbicides when they do work. A second factor is the loss of the soil's organic matter, down to an average of 1.5 percent in Ohio, when it should be at least 3 percent for efficient use of fertilizers, tractor power, and moisture-holding capacity, not to mention many other subtle influences on soil health, and therefore food quality. There are, for example, growing critical shortages of trace elements in Ohio soils—chiefly selenium, but also manganese, zinc, boron, and others, shortages that with less intensive farming and with maintenance of the soil's organic matter, were not previously problems. A third limiting factor to increased yields is the chemical residues building in the soil, though the full extent of their effects, harmful or beneficial, is not yet known. And finally, there comes a fourth and very apparent factor—the use of agricultural lands for various kinds of urban development. To compensate, new lands are brought into production on hills and drained swamps, and on the wide-open spaces of the arid plains, but these do not produce as well, or do so only with far greater costs in water and fertilizer.

But biological degradation stemming from economic pressure does not stop with the soil. All living things are affected, not the least of which are the farmers themselves and the communities in which they live. The quiet despair of farmers today is plainly due to the physical and emotional strain of economic pressures. Hypertension, depression, and alcoholism are now common among farmers. Marriages splinter, though more often than not the couples hide the disintegration behind a hollow façade of solidarity. Divorce is still not "acceptable" in rural circles, and therefore might be "bad for business." "The old widows who own the land, you see," says one farmer I spoke with, "might not sell or rent it to a divorced farmer." The new wealth creates not a feeling of security but paranoid fears. One farmer is afraid his wealth will become known and someone will kidnap his children. Another would not sign a controversial petition in his community even though he was in favor of it. Only when it became obvious that the majority of the people were signing it did he add his signature.

Says a farmwife, "All you reporters ever see is this hustling, coolheaded young farmer making big deals. Well, I'm married to one of them. He used to bolt down his supper and then go in the bathroom and throw up, he was so uptight. He got over that. Now he has to sit up all night sometimes because it hurts his gut so to lay down."

The whole élan of agribusiness has subtly ignored the humanity of the men who keep the dollars flowing into agribusiness's coffers. For example, when a writer filed his story about the farmer who had to sit upright through the night to ease the pain of ulcers, the farm magazine publishing the piece edited that part out. Farm magazines often portray the farmer as if he were a dehumanized extension of the computers and tractors they sell him. The farmer in their pages does not drink, sing songs, cuss, lose his temper, loaf, play, despair, or gape at pretty girls. He is a plasticized reincarnation of what some myopic public relations executive at John Deere or Dow Chemical hopes he is. He celebrates his coming of age by donning an FFA (Future Farmers of America) jacket and borrowing his first thousand dollars from a grandfatherly old banker. He spends his spare time during high school working like a slave and graduates owning forty acres and a $40,000 debt. Though he has apparently shown no interest in anything except his machinery, he appears in a few years on the covers of farm magazines with a wife (who is praised for running a house twelve hours a day and a tractor the other twelve), two toddlers, 500 acres, and a $350,000 debt. He disappears again into his cornfields, only to emerge triumphantly twenty years later with 2,000 acres, a million-dollar debt, and two sons who wish he'd retire and let them lose it all.

Alarmed by the lack of concern for biological realities, and the problems that this lack has brought to crop and livestock production, some farmers are trying to change. A motley group at best, ranging in size from large to mostly small, and calling themselves by a variety of names—organic farmers, ecofarmers, biological farmers, even biodynamic farmers—they agree only on a desire to cut back or discontinue the use of chemicals, fertilizers, and machinery they consider to be harmful to the life of the soil. At first, this "weird" group (there are still coffee shops in Ohio where the term "organic" is synonymous with "communistic") was the butt of ridicule because so many of them were rank amateurs or farmers known in their communities as not very good at farming of any kind.

But some of their practices have become more or less standard with all progressive farmers in the last few years, and there are indications that the movement toward a complete biological farming system may turn into a stampede. For example, Ohio is making strides in control of the alfalfa weevil, not with a toxic spray but by introducing a tiny predator wasp into the fields. There is a steady increase in the use of chisel plows and offset disks to replace the moldboard plow at least some of the time and thereby decrease soil erosion. There is even a small but noticeable shift back to putting legumes into crop rotations again both for fertility and weed control. Even the USDA has become aware of biological farming and is unlimbering its fossilized bones to the tune of a few costly studies, which, not surprisingly, are discovering what biological farmers have already learned: it works.

So who will put food on your table twenty years from now? This winter, Secretary of Agriculture Robert Bergland and his crew are holding meetings throughout rural America to listen to what farmers have to say about that question. Underpinning the meetings are several assumptions, none of which is proven. The first assumption is that government can do whatever needs to be done. The second is that American farmers need government help and that small "family farms" need it the most. The third assumption is that somehow the "new" biological methods are the answer to both the economic and the ecological problems of farming.

What Bergland is hearing about most often is the need for more small-scale farms and for curbs on large corporate farms which, because of inflation, can squeeze out smaller farmers in the competition for land. But there is good *and* bad in that view for Ohio agriculture. A reemphasis on small-scale farming is appropriate because small farms, *properly farmed,* can produce more food per acre at less cost in fossil fuel than large-scale farming, especially on hillier soil, where corn, soybeans, and "Florida rotation" can, and are, destroying soil through erosion. It is somewhat ironical that hilly Holmes County, farmed mostly with horses and Amish family labor, has greater cash receipts from farming ($43,138,000) than flat, rich Wyandot County ($38,300,000), which is dominated by the most advanced, large-scale farm technology in the world. You can be sure, too, that the greater portion of Holmes County's cash receipts is staying in Holmes County, while Wyandot's is flying away on the wings of national and international agribusiness companies.

The problem, however, is not as simple as choosing a "proper" scale of farming. Small is not beautiful unless small is skilled and dedicated. The hills of the eastern and south-central United States were destroyed for farming, not by large tractor farmers, but by small horse farmers a hundred years ago. The biological knowledge and skills necessary to make farming ecologically as well as economically sound are lost on most farmers today—big *and* small—and the laws that govern ecology are not very well known even by those scientists brave enough to call themselves ecologists.

In the process of learning what to do to allow this country to survive comfortably in the future, dividing farmers into economic groups makes little sense, and basing solutions on farm *size* makes even less sense. Instead, wise farmers must be discerned from the foolish, the skilled from the clumsy, and the knowledgeable from the ignorant. Some of the millionaire farmers need tutoring by the Amish of Holmes County. On the other hand, some of the small organic farmers who spurn big farmers' methods are ignoring knowledge they need. Blending the opposing forces of economics and ecology into a productive farm demands *supreme* skill. No scientist, no ecologist, no chemist, and God knows, no bureaucrat or social reformer is going to do it. An experienced, informed farmer is the only chance, and small family farm sentimentality to the contrary, *some* of the most experienced and best informed farmers are the large-scale ones. That's how they got there.

Herb Walton of Upper Sandusky is one of those farmers whose ingenuity is to be studied as well as admired. After World War II, when there were no tractors around large enough to suit him, Walton bought two army surplus tanks and converted them for farm work. When they didn't please him much either, he talked Bill and Ivan Schmidt at their shop in Lovell into building the first large, articulated-steering, four-wheel-drive tractor, the grand-pappy of the monster Steigers of today. That was in 1947. Walton was also the first farmer (according to farming tradition, anyway) to put a corn head on a grain combine, now the standard "picker-sheller" corn harvester of the corn belt.

The antitechnologists and the small-is-beautiful folk will demur, but Walton can teach them much they need to know. He also raises chickens, in addition to farming large acreages. He puts tons of manure back on the soil. He is a student of the agronomist William Albrecht, the godfather of the ecological farming movement. Five years ago, in 1975, when the biological farm movement was first stirring in Ohio, some of the promulgators held a quiet, unheralded meeting in a church basement in the tiny hamlet of Marseilles. The meeting was attended by only a few farmers, mostly of the visionary type or the faddists who tag after every new movement. Bill Kurfess from Perrysburg, a pioneer in ecological farming, did most of the talking that night. Besides Kurfess, there was hardly a farmer in the room whom the USDA would call "successful." Except for one—Herb Walton was there.

"What are you doing at a meeting like this, Herb?" someone asked.

"You never know where you might learn something," he

answered. "We've got to find better ways of treating the soil. Maybe these people have some of the answers."

Walton divined twenty years ahead of time what the 1970s would demand. Maybe he divines now what the 1990s will need.

At the other end of the spectrum is the small farm of Mr. and Mrs. James Kemp, near Jerusalem. On about 140 acres, the Kemps have a small herd of Holstein cows, which they have patiently bred into some of the finest dairy stock in the nation. Never milking more than thirty cows, and usually less, the Kemps have made a very profitable farm—their pedigreed stock is in demand nationwide—in strip-mining country, on land the USDA would consider too rough for good farming. Size has little bearing on their economic situation. Knowledge, skill, and a biological sensitivity toward animals have everything to do with it.

Ironically, while the government worries about small-acreage farmers and small farmers worry about big farmers and big farmers worry about the IRS, it may well be the middle-income farmers—the seemingly established, successful, larger family farm—who are in real trouble, especially those who, as Erven points out, lack sufficient land base for efficient full-time operation.

Hiram Drache, a large, first-class farmer and innovator like Walton in the Red River Valley of Minnesota, whose "second business" is teaching college history and writing books, is convinced by his experience and long years of study that farms will indeed keep getting larger. And the land for that expansion will have to come out of the hide of the middle-income farmer. Says Drache:

> . . . the progressive farmers are going to continue to use their financial leverage for future expansion, all opposition notwithstanding. Only a major change in the free enterprise

> system could alter the continued growth toward larger farmer-oriented, commercial farm units. On the opposite side, there is not likely to be much of a decline in total farm numbers because of a very strong desire for part-time farming by fully employed urban workers who prefer country living.

Drache's statement, made in 1975, is prophetic of what is happening, at least in Ohio. If so, the outcome is not all bad news for anyone, including ecologists who worry about the environment and political scientists who fear the land will all fall into the hands of a new, landed oligarchy that will destroy the democratic system. The part-time farmers are more than keeping pace in numbers with the high rollers, and they, more than any others, will be able to cooperate with new biological farming methods without fear of economic loss, until the new and necessary changes are perfected. This is not all bad news for the middle-income farmer, either. The changes will come gradually enough so that few will have to alter their lifestyles. Many have already taken second jobs or begun side businesses and enjoy the change of pace. Many will retire. Younger ones who have not done so will join the ranks of the part-timers and will probably, with their farming experience, make a better living from farming than they are now.

As for Robert Bergland's meetings, the message he's getting from farmers is a simple one: "Put your own financial affairs in order and ours will take care of themselves. Turn your scientists loose from political influence—your economists as well as your biologists—to work out ways a farmer can be rewarded for farming more ecologically. Go sit at the feet of the Amish, of the Kemps and Waltons, and be very quiet while you listen."

3

Our Hidden Wound

I'm a hayseed, I'm a hayseed,
and my ears are full of pigweed.
How they flop in stormy weather—
gosh oh hemlock, tough as leather . . .

—From a children's rhyme heard in the Midwest in the 1930s and forties.

1992

Most of us grew up in a society where *farmer* was often merely a synonym for *moron,* and I am quite sure that many farmers are still haunted by feelings of inferiority laid on them by this kind of urban and urbane prejudice. In fact, I suspect that many of the most competent farmers among us continue to expand their farm empires not out of greed or an insatiable desire for wealth, but because they feel compelled to prove again and again that, by God, they are not inferior to anyone. They want to cram that fact as far down the throats of their boyhood taunters as they can and, sadly, they spend their lives doing it.

In my high school days in the late forties, supercilious town girls routinely claimed that milking cows caused hands to grow too large and rough and the reason farmers had big feet was that they went barefoot too much. Lord help the girl who wore a print dress made from a grain sack, although the dresses were as pretty as any. A boy who came to school with chicken manure on his shoes, as could easily happen, or with the smell (real or imagined) of the cow stable on his clothes, instantly became an object of derision. Wearing bib overalls, which, ironically, are all the urban rage right now, brought automatic jeers, and after a while we refused to wear them, even at home. When the school lunch program came along, country children whose mothers packed a lunch for them, believing for some strange reason that parents, not the government, should feed their children, were restricted to a separate part of the lunchroom, and this separation soon carried with it a stigma not unlike the segregation of blacks in "their own place." Farm work was in all cases put down as "nigger work," and it was too bad, we were told, that redneck country kids were condemned to it. One of our textbooks, with all good intentions, I'm sure, had a chapter entitled "Farm Folk Are Human, Too." My mother, half-amused and half-dismayed, showed that page to my father. He took one look and hurled the book across the floor.

We farm kids came to school possessing intricate and valuable knowledge about manual arts, food production skills, and the ways of nature—all of which our urban counterparts desperately lacked, as is now apparent from the actions of well-meaning animal rightists and overzealous environmentalists; yet most of the teachers not only ignored this treasure trove of information, but belittled it as having no relevance to life. Kamyar Enshayan, of the Sustainable Agriculture Program at Ohio State University, calls this "paradigm negation" and says that rural students coming into the university are still treated as if what they have learned at home, from tradition or through farm experience, is of no importance. "This is, in fact, the way colonial powers always treat their colonies as a way of stripping them of their identity and destroying their independence," he says. "Farmers don't yet realize it, but rural areas have become no more than colonies from which cities are sucking the wealth."

In high school we accepted the urban prejudices against us in a solid, simmering silence that erupted into rebellion only once that I recall—a violent, bloody fistfight in the lobby of our local theater. The fight started when a "townie" called one of us a "clodhopper" once too often.

It wasn't so long ago, really, that that kind of prejudice was perpetuated all over America. We who are now in our forties and fifties bear the scars of these prejudices as part of what Wendell Berry, the poet and farmer, calls "our hidden wound" in his book by that title. And we know, like the blacks know, that the prejudice is far from gone: it has only become more slyly silken in its displays. Though the scars have healed, they ache whenever the cultural weather shifts.

Some farmers flaunt the prejudice by wearing dirty clothes to the bank to borrow a quarter of a million dollars. Others overcompensate by dressing up to look "respectable" for the banker. That's also why they get the car washed every time they're in town. Some want to be called "agribusinesspeople" rather than farmers even if it does take half an hour to get that word out. Almost all of us are suckers for the "urban counterpart" argument. Salespeople know that a good way to get a farmer to buy their product is to hint that it will enable us to live "more like your urban counterpart." Those who follow that allurement to its logical conclusion *become* urban counterparts, because it is patently impossible for a farmer to live like a city person.

How many generations does it take to heal the scars of prejudice completely? I wonder. I have a notion that prejudice is never eradicated, just transferred. When the "hillbillies" moved into our county from Kentucky during World War II, the focus of urban prejudice switched to them because they were even more "rural" than we were. Nursing our wounds, we farmers, who should have been sympathetic, joined with the townspeople in inflicting the wound on them. When the Mexican fieldworkers came, another segment of society colonized out of its own farm traditions, the "hillbillies" joined us, glad no longer to be at the bottom of the pecking order. Although there are hardly any blacks in the county, they are still referred to broadly as "niggers" by more than a few whites including most of the farmers; and "niggers" are still thought to be oversexed beyond control. I suspect, in fact, that farmers tend to hold on to such hoary racial prejudices in retaliation against their own hidden wound. Misery loves company.

Our county has just come through a nasty school consolidation fight in which, as usual, the bureaucracy won and the farmers lost. The school in the village of Harpster was closed (along with another township school). Being on the task force that undertook to study the matter, I was involved up to my ears (how they flop in stormy weather) in that battle. I had all the available figures pertinent to the school closing, and those figures did not show that there were any savings to be had by closing the Harpster school. Nor was there any proof that consolidating the schools meant better education. (In fact, nationally, more and more evidence points to quite the opposite conclusion.) Not even population decline could be cited as a reason for closing the Harpster school, because the area was gaining population. But argument was futile since the state of Ohio, like most states, is committed to consolidation. And latent in that policy is a contempt for rural people. Wayne Fuller, a professor of history at the University of Texas, has soundly documented this contempt in his recent book *The Old Country School.* In order to gain control of the independent school districts, professional educators undertook a campaign, beginning in the nineteenth century and intensifying in the twentieth, to discredit country schools in the eyes of state legislators. The professionals, often bluntly, said that farmers were too ignorant to be capable of running schools. Fuller points out that in most cases, the farmers' ideas about education turned out to be better than the professional educators', and that in following the latter's course, we now have a large percentage of our population that can't even read intelligently. My friend Craig Bowman, who with his sons farms about 4,000 acres today, was a leader in both of the futile fights to save Harpster's high school in 1960 and its elementary school in 1990. He nods when I tell him about Fuller's book. "One reason we lost those battles, especially in 1960, was that many farmers half-believed that those yahoos in the state education department knew more about what was good for their children than they did, and they wouldn't stand up to them. Of course. Society trained them that way."

Even in our rural county, teachers encourage students not to think of themselves as coming from Harpster, or Marseilles, or any of our little villages or townships, but from the Upper Sandusky School District, which is perceived as a nobler root from which to spring. "Big is better" is the myth behind the myth that country people are somehow second-rate. And that may be why farmers so readily embraced the slogan "Get big or get out."

But it is not necessary to blame education for the prejudice against farmers, since television, the real educating force in America, reinforces the myth with one prime-time show after another.

The bigotry is not even veiled. Night after night, one dramatic episode or another will follow the adventures of a character who just had to get out of a "backward" rural area in favor of the, tah-dah, City. Getting out of rural areas for fame and fortune persists as a story motif even though it flies utterly in the face of reality. The competent farmers and businesspeople who stayed in our county are at least as financially successful as their peers who went to the city, and they don't have to pay $300,000 for a $90,000 home, either. As one refugee back here from the big city says: "As for the cultural advantages of the city, who needs the traffic hassle? Electronics brings 'cultural advantages' to one's home, wherever it may be." (The "cultural advantages of the city" is another side of the prejudice against farmers. Why does no one speak of the cultural advantages of the country? For example, is a well groomed, ecologically kept, sustainably fertile farm any less cultural, any less artful, than paintings of fat angels on church ceilings?)

I am sure that the reason for the prejudice so many farmers exhibit against the Amish (the most biased like to infer, with a snigger, that Amish women are oversexed, like black people) is that their lifestyle unwittingly jabs at our hidden wound. The Amish remind us of ourselves fifty years ago, when we lived much like they do now and were ridiculed for it. And it is embarrassing to us that the Amish prove we could all make a decent living in farming by *not* trying to live like our urban counterparts.

What is so curious about the inanity of prejudice against farmers is that it exists right alongside the opposite prejudice: that farmers are the moral backbone of society. Farmers, of course (including the Amish), can be just as ornery as anyone else. This overly favorable image gains more credence the farther it is removed from agriculture. The wealthy townhouse dweller who has seldom been anywhere except Manhattan and Bermuda (and, as a result, is far more provincial than most farmers), thinks of the "man of the soil" as a kind of yeoman saint in overalls, working without surcease in the peace and quiet of God's country to feed the world. This image lasts until said townhouser builds a million-dollar home in the country and the farmer next door starts spreading manure. The age-old contempt quickly returns and many farmers who must try to "feed the world" next to suburbs are not even allowed to work in their fields after dark.

The prejudice against farmers carries far from the farm. A New York City magazine editor cannot keep from displaying just a tad of superiority when talking about the work of a farm writer like myself. Usually it is more than a tad. When a Camden, New Jersey, columnist reviewed my book about Andrew Wyeth, which I wrote in 1970 while I was an editor at *Farm Journal*, she wrote most kindly but expressed surprise that such writing could come from someone who worked on a farm magazine! We farm writers, nursing our wound, aid and abet that prejudice ourselves: invariably, when one of our associates leaves our ranks for work in another field of journalism, we say that he or she graduated to a higher rung on the ladder. Why is *Time* more important than *Farm Journal*? It is difficult for the urban mind to swallow the fact that a renowned poet and essayist like Wendell Berry, or an accomplished musician like Elmo Reed, is also a bona fide farmer.

This low opinion of our work causes many farmers to see their land as nothing more than a factory or mine or "resource" from which to extract money. They remain unaware of its exquisite beauty, its natural wonders, and its potential as a sanctuary for the recreation of the human spirit. They ignore its natural pleasures in favor of faraway vacation spots: the same farmer who gasps in awe at a redstart in Cuba (once it is pointed out to him) does not know that the same bird visits his Ohio farm every spring and fall. The farmer who destroys the wild sanctuaries of his own farm uses the money to hunt and fish in Canada. He dines lavishly in gourmet restaurants on food that is not nearly as "farm-fresh," "free-range," or "organically pure" as the meats and vegetables he could grow in his own backyard and barnyard. Eschewing the good life of his own farm, he eschews the good life of his own neighborhood. His barn is no longer full of laughing, romping children or grandchildren, his hillsides no longer echo the happy cries of sledders, his pond no longer draws the swimmers and ice skaters of his community. There is no community. The neighbors have all gone to the city. The village churches and schools and taverns and inns that once were scenes of far more delight than the boring, manufactured uniformity of tourism are boarded up.

If we farmers deny the magnificence of our own rurality, how can we blame urban society for treating us the same way?

4

The Failure of Agricultural Education

1992

The book that Ohio State University librarians say is one of their most consulted references is kept on special reserve behind a locked door. Why the door is locked is something of a mystery to the visitor, since as soon as he presses a button beside the door, as the sign bids him, a bell rings and a librarian immediately ushers him in, although she has never seen him before and does not know that he is a spy from the CFR—the Contrary Farmer's Revolution—of which he is the sole member. He asks for "that book that gives everyone's salary," and without ado, she pulls the two-volume set from a shelf and sets it before him. As Abraham Lincoln surely must have hoped when he signed the land grant university system into law in 1862, Ohio State remains one of "the people's colleges"—if you know which buttons to press.

The reason the Contrary Farmer is interested in professorial salaries is because he is, in his own words, damn mad. A professor recently told a group of farmers that, sad as it might be, more of them would have to face reality and go to town to get a job. At that moment, the Contrary Farmer declared war on the university. By God, he said, if farmers were going to have to face the reality of continuing land consolidation into fewer and bigger megafarms, then when were the professors going to embrace the free market principles they so avidly mouthed, get off the tax dole that protected them from reality, consolidate Ohio State, Purdue, and Michigan State into one big mega–ag college and tell two-thirds of their colleagues to "go to town and get a job." The Contrary Farmer knows deep in his heart that the university is not the root cause of rural society's decline, but he is not going to let the college walk away from a cultural tragedy for which the university, by abandoning moral and philosophical leadership, is as much to blame as anyone. Without its leadership role in a community, a university is just what Professor Bert Hornback says it is: "a corporate entity whose purpose is to raise money to pay outrageous salaries to hundreds of greedy administrators and nonteaching faculty." Disgusted with the system, Hornback recently resigned from the University of Michigan, where he was widely regarded as an excellent teacher.

Most College of Agriculture salaries at Ohio State are split between two and sometimes three different funding sources: the Columbus campus, the Ohio Agricultural Research and Development Center (OARDC) at Wooster, and the Cooperative Extension Service (CES), which is the organization charged with the job of disseminating to the public the agricultural and home economics information generated at the college. Unaccustomed to the rarified atmosphere of higher education, the Contrary Farmer almost swoons at some of the totals. As in the other colleges in the university, the research and teaching staffs at the College of Agriculture are divided into four levels, not counting department heads and administrators. In general, instructors make from $25,000 to $35,000; assistant professors, $35,000 to $50,000; associate professors, $50,000 to $70,000; and full professors, $70,000 to over $100,000. Tenure usually comes at the assistant or associate level. Counting CES personnel out in the county courthouses and staffs at various branch experiment stations around the state, there are about 2,200 faculty and staff. The dean of the college, Bobby Moser, oversees an operating budget of about $100 million for resident instruction, extension service, and research programs.

The programs this money and other monies from grants, gifts, and endowments make possible are numerous and varied—everything from breeding a new and improved variety of wheat to advising homeowners on how to adjust a microwave oven; from projects such as working with NASA on satellite sensors that can help the government and grain dealers keep an outer-space eye on crops worldwide, to a grant to Robert Gustafson, chair of the ag engineering department for "Technical Assistance to Moroccan Participant Driss Massahou on His Dissertation."

The college actually embraces two colleges, each with a research center. OARDC's farm and test plots outside Wooster are considered the main research center, but the Columbus campus still retains the remnants of a research farm too. Most of the once verdant cow pastures there are marked off for football parking and soccer fields nowadays, an apt commentary on what is happening. In addition to operating a research center, OARDC, which is ninety miles from Columbus, is also the home of a two-year technical college, the Agricultural Technical Institute. The Contrary Farmer asked an ATI staffer if it seemed strange to him that a College of Agriculture would have to spawn a second college that also teaches agriculture. "Well," he said, "there was some concern that the four-year programs at Columbus were not fulfilling the needs of students who really wanted to learn how to farm." Oh.

Ironically, most credits at ATI do not transfer to Ohio State, a situation that upsets ATI students who decide to go on to Ohio State, and which upsets some ATI professors, too. When the Contrary Farmer asked Dean Moser why the credits were not accepted, he said, "We're working on that," but chose not to elaborate. Rosemarie Rossetti, a professor in the ag education department, was franker. "There is still a notion that vocational education is inferior, not really on the same level with college work. But I keep arguing that all our education is vocational, whether we're training lawyers, chemists, farmers, or concrete-block layers."

The spread between OSU salaries and those at private colleges is wider than any stretch of a raised eyebrow. Professors at small colleges like Heidelberg, in Tiffin, so the Contrary Farmer is told, are lucky to make $35,000 carrying a full load of classes. Sister Christine Pratt, a nun who is president of the National Catholic Rural Life Directors Association, says that that's about the top for professors at Catholic colleges, too. "Less than that if it's a women's college," she adds dryly. Asked if he made $100,000 a year, David Orr, head of the Environmental Studies Program at Oberlin College, replied, "No, I do not, and I know of no reason why any professor should. Ag economists might be paid like Chinese doctors—peg salary to the health of the rural economy."

The Contrary Farmer asked Dean Moser how he could justify

paying his professors so much more than private colleges pay. First he said that he was not so sure that that was the case. Assured that it was, he said he had to compete with other tax-funded Big Ten schools and private industry for the best people. The Contrary Farmer said he was tired of hearing that answer—that he doubted mightily that most of the highly paid professors could get from private industry the kind of money they were being paid now. The Dean stared at him. The Contrary Farmer knew that stare: he had seen it years ago from another dean when, as a student, he had tried to argue that graduation ceremonies were stupid. Hush your mouth, you naughty boy.

On top of extravagant salaries, professors enjoy a gold-plated benefit-and-retirement program that the taxpayer foots a healthy part of. The professors also enjoy very generous health coverage from the university.

But the Contrary Farmer would never have declared war on this mystifying world of self-aggrandizing noneducation were it not for the underlying irony of the situation. As professors go on getting raises while "serving" agriculture, the number of Ohio farmers keeps dropping, and so does the number of students in agriculture. Ohio lost 4,000 farmers in 1990, more than any other state, and is now down to about 75,000. *Ohio 21,* an OSU ag college publication, points out in its March 1991 issue that undergraduate enrollment dropped 42 percent between 1980 and 1990. Milton Friedman's "law of bureaucracy" is in effect: as output falls, input increases. Furthermore, many agricultural research conclusions, although touted as objective science, change regularly as new "facts" are discovered. And much of the economic prognostication upon which agricultural education policy is predicated turns out to be simply wrong. The whole organic farming movement was completely missed by the land grant colleges.

The reality of a rural society in decline and a shaky agriculture that has to be propped up with almost constant subsidy programs are, in fact, bringing a chorus of angry criticism upon the university—many more voices than that of the Contrary Farmer. By the terms of the Morrill Act of 1862, Congress specifically prescribed that the land grant colleges (created through the grant of thousands of acres of public land to each state's congressional delegations) were intended "to promote a sound and prosperous agriculture and rural life as indispensable to . . . national prosperity and security." Critics contend that the directive has been corrupted to mean: to promote a prosperous oligarchy of wealthy absentee landowners, megafarms, and international agribusiness firms as indispensable to a global, centralized, urban power structure.

Why have the land grant colleges not only stood by silently but also actively encouraged the trend by consistently espousing technology and tax policies that guarantee the destruction of moderate-sized family farms? When the Contrary Farmer decided to brave the storm he would surely arouse (criticizing the universities is tantamount to criticizing churches), he was amazed, almost to dumbness, at what he found.

Only one OSU professor has published research reflecting concern for the decline of rural society. Linda Libao in her new book *Locality and Inequality* gives careful and almost tediously thorough evidence that the concern for the decline of commercial family farms is justified. Libao, a sociologist, points out that we are now losing not just "fringe" farms but mainstream commercial family farms, and with them local supporting industries. The result: counties dominated by corporate agribusiness farming tend to have slightly higher levels of unemployment and poverty and lower levels of income than counties in which full-time family farms are dominant. "The bad effects of corporate farming aren't huge," she says, "but corporate farms are definitely not as good for the community as family farms."

But most professorial reaction to such news is a hopeless or helpless shrug. Kismet. Land grant college professors believe that what is happening is the result of some iron rule of economic determinism to which we must all bow (except, of course, highly paid administrators). "We should not be overly concerned," says OSU economist Joseph Havlichek. "Many of these people are not going completely out of farming, just partially. I don't mean there aren't some very sad cases, but many [of those] getting out of farming will be better off for it." Fred Miller, chair of the agronomy department (salary, $102,000), who is often dismayed by the Contrary Farmer (Miller and Havlichek are kindly men—in fact, everyone in the College of Agriculture is extremely nice—trying to argue with them is like trying to punch a great big foam ball), always responds patiently: "For you, me, OSU, or others to buck this economic force through appeals to stewardship and ethics in favor of a system that does not use cheaper resources to compensate for the high cost of land and labor, or to try to resurrect rural America through agrarian-generated wealth is an exercise in futility." To which the Contrary Farmer barks back that the substitute resources aren't cheaper in the long run, and he is not trying to resurrect anything; he only wants some way to keep megafarmers from continuing to outbid smaller, family farmers when land comes up for sale.

In addition to their faith in the god of economic determinism, the professors seem to nurse a childlike trust in a science that will always come to the rescue. When the Contrary Farmer tries to talk about ethics with Dean Moser, Moser says, "Science is what we're involved in."

"You mean you do not have to concern yourselves with the ethical ramifications of your science?"

"Don't you put words in my mouth. I'm saying society decides the ethical standards. We just supply scientific facts."

Says Miller: "Sure, fossil-fuel-based systems can't be sustained, but until the scarcity of this resource is reflected in increased prices and the need to develop alternative fuels in agriculture, there's no incentive to change. We will change when prices dictate, not when authors of books and philosophers demand it should change." The Contrary Farmer, who has been known to author a book now and then and to philosophize most of the time, barks again. Okay, so out of practicality, let the technology of monopoly continue to carry us toward consolidation until, as history always proves, the consolidation collapses of its own weight, like the USSR did, as GM is apparently doing now, and as the public school system is inevitably going to do. Are humans so stupid that they are doomed to a constant gyration between consolidation and collapse in the name of short-term profit and so-called scientific fact, which changes every ten years? Is it not the role of the university to stand adamantly against this lunacy instead of getting in the hog trough for its share of the short-term loot?

Some professors deny there is a problem. Privately one of them responded to the fomentations of the Contrary Farmer and others by pointing out that some farmers without debt were doing well, that they, too, were bidding up the price of land, that they were still producing a surplus, all of which was supposed to indicate that while philosophers (a word often used as a derogative by those who themselves hold doctor of philosophy degrees) can challenge agricultural structure, within the free-market structure agriculture is doing very well.

Hearing what he believes to be a monstrous untruth, the

Contrary Farmer can control his wrath only by remembering that there are economists elsewhere, like Harold Breimyer and his colleagues at the University of Missouri, who have spent their careers pointing out the fallacies of that view. Breimyer once told the Contrary Farmer, "The forces that affect our structure of agriculture are neither 'free' nor 'market.' The decisions of who will control agriculture—and *control* is the key word—and what form agriculture takes cannot be made strictly on economic grounds." At the end of his career, frustrated over the economic policy our government has pursued, he confided, "One can wonder if the system is rigged against family farms."

Kamyar Enshayan, who often disagrees publicly with the prevailing complacency at OSU, gave an impassioned speech to the disgruntled farmers at the Ohio Farm Crisis Rally in September of 1991, denouncing the free market concept. "Free market is a myth," he said. "It implies the prevailing notion that there is some fair, value-free, invisible hand which automatically allocates resources in the best and most efficient ways. But as far as I know, there is no such thing anywhere, except maybe at the very local level, as in a small farmer's market."

Wendell Berry, a working farmer, poet, and professor of English at the University of Kentucky, and the critic who most provokes the ag professors at land grant schools, skewers the hypocrisy he detects in professors who sing the praises of the free market while living in luxury off the taxpayers. He tells the Contrary Farmer (as he wrote earlier in his book *What Are People For?*): "The land grant universities, in espousing the economic determinism of the industrialists, have caught themselves in a logical absurdity that they may finally discover to be dangerous to themselves. If competitiveness is the economic norm, then how can these institutions justify public support? Why should the public be willing to permit a corporation to profit privately from research that has been subsidized publicly? Why should not the industries be required to do their own research, and why should not the laws of competition and the free market, if indeed they perform as advertised, enable industries to do their own research a great deal more cheaply than the universities can do it?"

Sister Christine Pratt has a master's degree in rural sociology from OSU. Back in the eighties, during the worst days of the farm decline, she once climbed up into a combine cab and talked a farmer with a gun out of killing himself. She minces no words in her criticism of the university: "When I became interested in rural ministry, I didn't know much about agriculture, having been raised in the city. So I went to OSU, and I'm glad I did. I learned the system and learned I didn't like it much. A few months ago I went to a marketing session being offered by the Extension Service. One of the Extension fellows asked me why I was there, as if I didn't need to know anything about economics. 'Aren't you that nun who goes around holding failed farmers' hands,' he asked me. That's the kind of put-down I have to deal with lots of times. It irks me now, as I work with families who are in despair over losing their farms, to hear Tweeten and Company [Luther Tweeten is a university ag economist] going around the state giving speeches that always begin with a kind of disclaimer—'Yeah, we know that farmers are human beings, but we have to consider farming as a cold, hard business.' To me that's hypocrisy. They don't seem to extend that cold, hard business attitude to their own tax-paid salaries. Besides, we're not here to serve economics, but to make economics serve people. I'm not sentimental about the past, as I'm accused of. I'm for technology, but I want technology for life, not death. Land ownership is foundational to justice. Industrialized agriculture takes land away from people, but it can hold it only with cheap labor, which is exactly what is happening. Justice always suffers in this situation."

She pauses, then thinks of another point. "Sure, some farmers who go broke have made bad business decisions or aren't very competent in a business way. But often it is just a matter of timing. They do what the successful guys do, but at the wrong turn of the market or of the weather, which are unpredictable. And some of them were trying to follow instructions from OSU when they went under. I have noticed that failed farmers as a rule have excellent nurturing instincts and great sensitivity for environmental problems, but there is little room for that in the ruthless arena of competitive agribusiness."

Did she have any professors at Ohio State who presented her viewpoint in class? "No," she replies. "Actually there was so much competition and backbiting among staff members that anyone who did agree with this philosophy was afraid to say so out loud."

David Orr, from Oberlin, has gone so far as to suggest in a speech before students and faculty at OSU that food production is too important to be entrusted entirely to agribusiness experts. "I think we might better achieve good sustainable farming if we approached it as a liberal art with technical aspects rather than as a series of narrowly focused technical specializations. The latter approach too often leads to bad farming.

"We have to reconceptualize agriculture," he continued, "and face up to the realization that you can't maximize profit in farming without minimizing environmental well-being. That lesson should begin in the university. Students sealed off from reality on college campuses or in the world of computer screens and malls don't learn much about natural systems, don't learn how to care about the living world, or about each other. They and their professors specialize in specific narrow areas of knowledge and don't learn how to make connections with other areas of knowledge. So they can't appreciate, for example, the idea that the world may not be able to endure the completely urban civilization that society is now trying to form."

Some of the criticism leveled against the College of Agriculture has little to do with the agricultural situation itself but is part of a growing discontent with publicly funded universities everywhere. Bert Hornback, speaking of the University of Michigan, sums up the general criticism. "Those who run it—and many of those who work in it—are no longer interested in teaching, and they care not the least about students. They are interested in what is called 'research' instead. Students are tolerated because they pay tuition. And the purpose of the research is to bring in money to further careers. . . . The university doesn't just steal money from its students; it also gets, and wastes, millions of dollars of taxpayers' money. In times like these, when so many people are homeless and hungry, it is immoral for the university to use state and federal tax money the way it does."

Enshayan insists that there is a great reservoir of talent at Ohio State that needs only to be redirected, yet he believes that the same criticism applies to that university as well. "In spite of much lip service to the contrary, there's little reward at Ohio State for good teaching, as compared to research and writing," he says. "It's assumed that if you are a good researcher, you will be a good teacher. Even those who want to work at being good teachers complain that so much emphasis is placed on research and publishing that they don't have the time to devote themselves properly to teaching. Two of the only three really good professors I had as a student here were not tenured and left."

Confronted with this accusation by the Contrary Farmer, Dean Moser almost showed emotion for the first time and flatly denied that students were not given special and close attention in the College of Agriculture. "I can't speak for other colleges, but this is something we pride ourselves in and I intend to make it an integral part of my term as dean. I will continue to initiate programs that will reward good teachers—and where it counts, too, in salary."

Whatever the truth of this statement, it is hard—at least for an outsider—not to conclude that much of the research undertaken by professors is self-serving and without redeeming social value. For example, in both 1983 and 1984, Lawrence H. Newcomb, associate dean of agriculture and director of academic affairs (salary $102,900) received two grants, one of $7,300 and another of $7,917 (chicken feed by university standards) to do an "Analysis of Burnout Among Faculty in Agriculture." For the years 1985 through 1989, he received $35,760 for an "Assessment of Levels of Cognition of Instruction in the College of Agriculture."

Keith Smith, associate dean and director of the Ohio Cooperative Extension Service (salary before recent appointment as director, $88,032), received grants of $53,280 and $23,240 in 1988 to fund a "Learning Through Writing Project." Along with two other colleagues, Smith received, also in 1988, a $16,000 grant for a "Study of Stress and Turnover Among Extension Directors and Administrators."

Perhaps the most ludicrous study funded in the ag education department was done in 1987–88 by Larry Miller. It was: "Perceptions of the Future Farmers of America as Held by Significant Others."

"With few exceptions," says Enshayan, "much of the educational research in agriculture and many other fields at Ohio State is based on quantitative analysis, ignoring much of the inherently unquantifiable, qualitative aspects of human life." "Farmers and rural people are simply 'data' or 'vocational groups' and the rest of us 'consumer units.' These educators think they can reduce the depth and diversity of human interactions and experiences to a bunch of numbers. They think that makes their research look more 'scientific' and therefore more respectable. I have a friend who says they suffer from 'slide-rule envy.' "

On top of the criticism and blame, which, needless to say, most professors feel is unwarranted, the College of Agriculture has been hit by budget cuts in addition to slumping enrollment. The one abets the other. However, says Rosemarie Rossetti, "Not all of the slump in enrollment can be credited to lack of student interest. As part of the budget-cutting process, the university has decided to limit freshmen enrollment, go after a more elite student, and raise tuition. This hits the College of Agriculture especially hard. Not only do rural kids have less money in general, but it is a common fact that they average lower in ACT scores. So it would appear that fewer of them will be coming to Ohio State. This policy seems to me in direct conflict with the concept of the land grant college."

"We got hit with a double whammy," says Havlichek. "We had already been downsizing for about four years, and now comes this across-the-board cut to all the colleges in the university. We're cut to the bone now."

Miller, displaying just a wee bit of impatience at the Contrary Farmer's challenge to merge Ohio State with Purdue and Michigan State into a mega–ag college, replies, "We've been struggling with these issues for two decades. We *are* economizing, consolidating, cutting back where we can. You know, we have to meet accreditation standards and if we limit curriculum offerings too much, accreditation is forfeited. We've been downsizing through retirements and other departures since 1989. Extension downsized itself in the late 1980s. We've made dramatic shifts in the agronomy department in focus and priorities, including environmental science and molecular biology, and done it with fewer faculty. We do share resources with Purdue, Penn State, and Michigan State. The University of Kentucky services much of our tobacco extension needs in southern Ohio."

"We had a deficit of $7 million at OARDC before this last cut came along," says Moser, once more emotionless and matter-of-fact. "Now we have to cut six percent more from our annual budget, and take that all out of the second half of the year, so it's really like a twelve-percent cut. We're cutting back on operational budgets and not filling positions."

Through all this poor-mouthing, the Contrary Farmer is waiting to hear someone suggest cuts in salary, but no one ever does.

To counteract the decline in enrollment, the college is making a strong pitch to urban students and offering more courses in biotechnology, food science, environmental science, and other "nontraditional and nonproduction" fields of agriculture. Critics say these subjects would be better taught in other colleges, such as the College of Biological Sciences, because ag researchers are too biased in favor of farm methods that are harmful to the environment. "This is an attitude we have always had to fight," says Havlichek. "There's always been a myth that somehow agriculture is a little bit second-rate. This college is as damned scientific as any other college in this university. Our economics department is respected worldwide. Serving production agriculture is what we will always do, but that is really only a small part of what we do. We need to expand into other areas faster, or quit. I want to change our name to Food, Agriculture, and Environmental Science." (Penn State has already changed the name of its College of Agriculture to College of Agricultural Sciences.)

The "forsaking of agriculture," as the Contrary Farmer calls it, is already evident throughout the college. The Cooperative Extension Service, for example, has, as one of its stated priorities, to "develop strategies for retaining and expanding industrial firms in communities with the goal of absorbing displaced farmers." Extension agents are being retreaded to serve as "economic development agents" to hasten the assumed day of rural industrialization. Now the Extension Service has received a major grant to start up a program that would educate rural people about low-level nuclear waste dumps, so that, as Moser tried to explain to an infuriated Contrary Farmer, the people can make up their own minds about whether or not they want such a dump in their county.

"You mean that you believe a rural community needs to be told by the Extension Service why they don't want a nuclear dump in their county?" the Contrary Farmer asked in utter disbelief.

"Well, they need the facts to decide why they don't want it."

"There's no community in the world that is going to accept a nuclear dump willingly, and they don't need the Extension Service to make up their minds," the Contrary Farmer replied and then, dumbfounded, heard Moser say:

"People need to look at both sides. It could mean a lot of jobs, you know."

And so, on to the final forsaking of farming: once the farmers have been displaced, the land will be ready for use as a dumping ground, so the poor, dumb sonsabitches can have jobs.

A look at the trustees of Ohio State suggests why university policy might tilt toward the industrialization of rural countrysides rather than the fostering of a strong rural society. The story goes, as related in *The Ohio Farmer,* that when President Gee introduced Bobby Moser as the new vice president and dean of the college, one board member replied, "Not so fast. We have to approve him first." What is most ironical about that remark is that not one of the trustees, least of all the three currently on the agricultural affairs committee, is a practicing farmer. John Kessler is a developer who turns farmland into suburbs; he owns the New Albany Company. He is also a director of Banc One Corporation, as well as of the Limited, Inc., and Rax Restaurants. Ted Celeste, the chairman of the agricultural affairs committee, is founder, chairman, and president of Advanced Interactive Video, Inc.; he is also president of Celeste and Santer, Inc., a residential real estate brokerage firm. Michael Colley, the third member of the ag affairs committee, is a lawyer. It seems highly likely that such businesspeople might look favorably upon industrializing or suburbanizing rural areas and sending displaced farmers to factories, rather than fostering communities of small, tightfisted, independent farmers on the land who almost invariably vote no on taxes. Other board members are investment brokers, lawyers, a clothing tycoon who has given millions to the university, the president of a communications firm, and two students, who, according to the accepted campus wisdom, are merely a token presence on the board. Four of the eleven are lawyers, four are bank directors, and four are in real estate development or investment finance. Where are the farmers, whom the College of Agriculture supposedly serves? Where, in fact, are any lower- to middle-class representatives?

Another example of the shift in emphasis away from farming

can be seen in the agronomy department. Several recreational turf specialists (concerned with golf courses and suburban lawns) are now on staff, but not one forage specialist, a fact that would have been unthinkable just a few years ago. "But we are going to hire a forage specialist," says Moser. "That's in the budget."

"Yes, we are heavy into turf," says Miller, "because half our undergraduate students are in our turf program. That's what they want. Lawn care and the recreational turf industry is over a two-billion-dollar-a-year industry and these people pay taxes too and want their slice of the pie. The landscape business is equally important. Should our plant science efforts be concerned only with agricultural crops and stop at municipal borders? Likewise, should our Veterinary School not serve the urban cat and dog population?

"More and more of what we did in the past will be given up," he continues, "and in all parts of the university. There will be screams of protest when we do. There will be more programs supported by endowments or direct income. I'm working on an endowment for the agronomy department now. This trend puts us at risk to criticism of being in bed with too many special interest groups. But the fact remains that we, and most other land grant institutions, have passed from being state-supported to state-assisted institutions. Our faculty has been turned into entrepreneurs, 'grant- and gift-getters.' "

Although he does not believe taxpayers in general will vote to approve taxes for golf course managers, the Contrary Farmer now finds that empathy for the professors' stressful lot is replacing his anger. The Bottom-Line Demon that has turned the natural world of farming into a chemically soaked frenzy for money has also infected the formerly quiet "halls of knowledge" where dedicated doctors of philosophy were once satisfied, on modest income, to pursue truth. Now they have become "entrepreneurs," lurking like buzzards at a roadside, waiting to see where big government and big industry will dump their next rounds of funding and then competing against each other for the grants. The stress, the constant pressure to find more money, haunts everyone. Bobby Moser told the Contrary Farmer he had been on two retreats in January. Intrigued, the Contrary Farmer asked why. "We have to get away from the university for some quiet time to think," he was told. The Contrary Farmer can remember a time when that's what the university was for.

The College of Agriculture has made up its mind about what the future of farming is going to look like. The name of its magazine, *Ohio 21,* proclaims that confidence. Fred Hutchinson, a former dean, stated its mission in the first issue: "The title, *Ohio 21,* captures the thrust of our major initiative; to have educational programs that will position Ohio agriculture to effectively deal with the challenges of the twenty-first century." The Contrary Farmer asked Moser to describe what farming would be like in the twenty-first century. He would not fall for that trap. "I don't know," he said.

"Well, then, how can you position yourself for it?"

Again the naughty boy stare. "We have to try as best we can to plan ahead," he said.

But the plans are all based on a paradigm of ultramodern biotechnology and automation that assumes more and more consolidation of farms, of factories, of government, of schools, of churches, of everything. Meanwhile, all the actual economic signs, from Russia to General Motors, point to the probability that the consolidation craze has about run its course. What if Ohio State is preparing itself for a future that never comes?

A cross-departmental group in the university, under the Sustainable Agriculture Program, has begun its own publication, one that is in many ways the exact opposite of *Ohio 21,* and which does indeed address that question. *Sustainable Agriculture* provides information by and for small-scale and ecologically minded farmers. It also provides a forum for critics of the agricultural colleges and agripolitics. Response from the professorial ranks has been mostly negative, demanding that the newsletter carry a disclaimer that it does not represent the views of the university. Kamyar Enshayan, who is responsible for most of the editing, has added the disclaimer, though he protests, "I don't see such a disclaimer in any of the Extension bulletins or any other OSU publication. Is the Extension view the official view? Is the *Farm Outlook Guide* from the economics department the official view? I sure hope not. There shouldn't be any official view."

The Contrary Farmer goes home now more disturbed than ever. The university is in the same kind of trouble that farming is in, yet it builds a $14 million indoor football practice field for its athletes across from the ag campus, on land where the university dairy herd once grazed. When the Contrary Farmer bemoans the fate of the farm hen flock, as dramatized by the row upon row of old chicken coops moldering away behind the poultry building, Fred Miller says he assumes that the eggs he eats come from the huge Croton egg factory northeast of Columbus. He believes the eggs are just as nutritious as any farm-fresh, free-range eggs and probably cost less to produce. The Contrary Farmer says no economist has compared Croton's cost for egg production with his own hens, so it is not very scientific, as the professors always profess to be, simply to *assume* Croton is cheaper. Croton's egg factory, say OSU researchers, puts out the sewage waste equal to that of a city of 700,000 people. How much of that cost is figured into the long-range egg price? When Croton was fined in its early days for polluting runoff water, the fine went to the College of Agriculture to help Croton solve its problem. Instead of a fine, it was a donation! The university, as part of positioning itself for the twenty-first century, subscribes to Crotonesque visions of the future, and so far that has proved to be a self-fulfilling prophecy. But if tax policies did not favor giantism, what would the twenty-first-century farm look like?

No one in the university debates such matters (maybe Croton will pollute again and we'll get another grant?). The halls of knowledge have become havens for golf-turf science (even though both Chemlawn and Scott Seed Company, almost within shouting distance of the university, are doing that work better). Tenure is unnecessary anymore, since no well-paid professor says anything that requires that kind of protection.

5
Traditional Farming

1984

During the summer, my chickens get most of their protein supplement by eating blood-engorged flies off the cows. When I let the hens out in the morning, they scurry off to where the two cows lie in the shade, first attacking the big, slow horseflies they can reach from the ground, sometimes jumping up on the reclining cows in their eagerness. When the horsefly population is depleted momentarily, the hens linger around the cows' heads, nabbing the nimbler face flies. The cows never budge during any of these maneuvers, obviously aware not only of the benefits accruing to them but also trusting completely a chicken's ability to peck within an eyelash of their eyeballs without injuring them. This symbiotic relationship between cow and chicken must be common, or must have been common, on the traditional farm where animals often shared the same quarters. I remember hens on a Minnesota farm roosting on the cows' backs when the temperature sank below zero, but I have never seen any reference before to the cow-horsefly-chicken connection. Having noticed it, I tried to become more consciously aware of the humdrum barnyard activity on my own place, where I attempt to reproduce the traditional farm of about 1940 on a very small scale—a farm of thirty acres. (I use the word *traditional* to refer to those practices of mixed livestock, crop, garden, and orchard farming brought over from Europe and adapted, over generations, to middle-American climates and soils. There are other kinds of traditional farming, of course, such as that practiced by the southwest Indians.) The first observation I made—one I had hitherto taken for granted—was that the fly population on a cow tended to increase as summer progressed, but only so far, leveling off even though there was still more cow to bite and more blood to draw out. What halted the continued population growth? The chickens may have been part of the answer, but obviously only a small part, since they could catch significant numbers of the slower horseflies only. And the fly population tends to level off on traditional farms even when chickens do not take a direct interest in the process. This leveling characteristic was certainly not true on the confined dairy where I once milked a hundred cows daily. The cows were never let out to pasture there, and the fly population continued to explode until the frost, although we fogged the barnlots frequently with insecticides. The backs of my two cows became a frontier for new discoveries. One day a yellow jacket (ground wasp) buzzed up to a cow and after several unsuccessful attempts, managed to capture a fly and carry it away. I searched books for corroborating evidence but found none. Nor did I see that phenomenon repeated. I yearned for more time just to stand and watch. For two weeks in August, a flock of cowbirds descended daily upon the cows and sheep. Through binoculars I watched them dart after flies on the cows' backs. They also marched along on foot right beside the cows' noses, not only interested in face flies and other insects stirred up by the grazing animals but also, as far as I could tell, eating bits of the saliva-flecked grass dribbling from the edges of the cows' mouths. A mystery. But I was reminded of a report issued by the University of Minnesota a few years ago revealing that a chemical released by the grazing animal's saliva actually seemed to make the grass grow back faster than grass that was cut mechanically. I have seen no follow-up to that report. Knowledge of such ephemeral activity does not program easily in computers, I assume.

I also became a ponderer of cow droppings. When the chickens tired of chasing flies, they attacked the cows' manure—what we call cowpies—a normal occupation of chickens and hogs, but one which I never before took the time to observe closely. The hens tore the cowpies apart, reducing their bulk by a third, scattering the fecal matter over a broader area so that the grass was not smothered under it. From the cowpies, seemingly as delectable to chickens as apple pie is to humans, the hens ate partially digested grains and greedily gobbled chunks of digested grass. The hens were not starved to this diet. They had plenty of fresh grain and grass available to them. They preferred this food. In addition, they pecked at tiny specks of indiscernible stuff and fly eggs, plus other insects drawn to the manure, not to mention earthworms that had worked up from below to dine on the succulent organic matter. Over the course of the summer, I counted seventeen distinct species of insects in and about the pasture cowpies, all working out their life cycles in some kind of tenuous connection with the manure.

This last discovery may have explained another phenomenon drawn into the encircling relationships between farm and nature. The preceding spring, snow lingered into March and I worried that the returning bluebirds would have nothing to eat. During the first thawing days, however, I saw them flying from fencepost to ground and back to post again, just as they do in summer when feeding on insects. I examined the bare ground and the meltwater pools between the patches of snow. To my surprise, the scum of the meltwater was dotted with drowned bugs. Where they had come from so early in the year I cannot say, but they resembled closely the gnatlike insects that flocked around the cowpies in the fall, cowpies that in March were miniature islands in the meltwater pools of the pasture.

Awakened to any possibility, I turned my attention to the pair of woodcocks that persists in nesting at the edge of the pasture where trees crowd out from an adjacent woodlot. Woodcocks are fairly rare in this area and terribly shy. Why would they put up with my daily comings and goings and the cows tramping their shady territory to bare earth? I found the answer in an ornithology report. According to the report, cows encourage the presence of woodcocks: they tramp the ground under trees bare, aiding the woodcocks' search for earthworms, their principal food; cow droppings further encourage earthworms to the soil surface to become easy pickings for the probing birds.

The one creature notably absent from the pasture cowpies was the dung beetle. When I was a child, dung beetles—we called them tumble bugs—were a common diversion. I watched them by the hour as they fashioned their marble-sized balls of manure from the cow and sheep droppings, rolled the balls laboriously down the sheep paths, and then buried them in the pasture grass, each ball with an egg inside it. But around 1950 the dung beetles all disappeared.

In recent years, researchers at the University of California and at the U.S. Department of Agriculture station at College Station, Texas, have been introducing dung beetles to those two states. In two days, they say, a pair of dung beetles can bury a cowpie. Since our 200 million cows drop cowpies to temporarily cover an estimated 8 million acres of grass each year, the value of the beetles is enormous: they remove this covering before the grass under it is smothered and bury it before its nutrients are lost by leaching. At the same time, the prompt removal helps control livestock pests, especially face flies and horn flies, which lay eggs in the manure. The latter cost cattlemen $5 million a year in California alone, say researchers; and pinkeye, which is spread by a bacteria carried by face flies, causes another $1.5 million worth of damage each year.

In my part of the country, northern Ohio, the face fly appeared in problematical numbers almost exactly with the demise of the dung beetles. But what caused the demise? No one seems to know. It is easy to blame the careless use of insecticides, but more than likely the cause was the abandonment of traditional farming methods. The dung beetle disappeared at the same time that thousands of small-farm sheep flocks were sold and when cows and hogs were taken off pastures in favor of confinement; the permanent pastures were then plowed for corn. The ancient Egyptians made golden images of their dung beetle, the scarab, and placed them in the tombs of their loved ones—a symbol of the return of all that decays to life again. It is a symbol, too, of the basic tenet of traditional farming.

Other interrelationships were yawing off around my cowpies

in intersecting orbits. Where the manure decayed in the pasture, white clover sprang up as if by magic. Where the clover grew, the bluegrass followed in a year or two, as surely as night follows day. The grass greedily usurped the nitrogen that clover and rhizobia bacteria, working together, had drawn into the earth from the tons of free nitrogen in the air above each acre. The bluegrass crowded out the clover until the nitrogen was depleted. Then the clover came back to manufacture more nitrogen; the cycle continued as long as grazing and an occasional mowing controlled weeds and brush.

Upon these pastures, the livestock tied together other encircling food webs. They ate from a wide menu—grasses and clovers of various sorts, in addition to the bluegrass and white clover, and more than twenty weed species. They occasionally reached up and grabbed mouthfuls of tree leaves. They pawed and licked the virgin soil of the creek bank, even though they had access to a mineral block that, according to the experts, contains every trace element an animal needs. When certain weeds and grasses went to seed—particularly orchard grass, timothy, Kentucky fescue, and pigweed (amaranth)—the cows and horses took a sudden interest in eating the seedheads, knowing instinctively, I suppose, that these seeds contained more protein than the grains I might otherwise have had to feed them. And what the cows didn't eat, the sheep did. The traditional rule of thumb is that a pasture that will carry twenty cows will carry thirty sheep more and never know the difference. Forty sheep, says sheep- and cattleman Russell Conklin in Kentucky.

All the while, pasturing animals spread their own manure, to keep the whole rich process revolving. John Vogelsburg, on his large traditional farm in Kansas, has carried this idea a step further back in tradition. He has returned to stacking loose hay in his pastures. The cows feed themselves on it in winter, and by moving the location of the stacks each year, Vogelsburg lets the animals spread all their manure instead of having to haul it from the barn himself. He says that with the old stacking equipment, making hay is no harder than with a baler; the old method also makes a better quality of hay because hay cures better in an outdoor stack than in a barn loft.

The encircling food webs spinning out from the cowpies entwine the animals finally with the humans who eat them. Traditional farmers put chunks of sod from rich permanent pastures into the pens of sows and new pigs, knowing the animals will derive needed iron from the dirt. The iron in the animal eventually becomes iron in the human. Natural antibiotics live in healthy soil. My dairy calves, pasture-born, pasture-raised, and mother-nursed, have never gotten scours. Confinement cattle, on the other hand, must be pumped full of antibiotics to keep them healthy, resulting in the danger of bacteria becoming immune to antibiotics. The enveloping food webs will either produce healthy humans or, if we fail to see the connections, diseased humans. One of our new egg customers called frantically after she cracked the first of our eggs. She thought we would want to know that the egg's yolk was "terribly orangish." She had never before seen a healthy egg.

Not the least significant aspect of the interrelationships between traditional farming and nature is that much of the activity leading to production of food—the purpose of agriculture—proceeds without the expenditure of energy on the part of humans or machines. What machine, however electronically clever, can duplicate the accomplishments of a mere cowpie? Swedish scientist Staffan Delin has recently theorized that "it well may be that the biological processes are many magnitudes of order more efficient than the industrial ones." I suspect that this insight—call it *biological efficiency*—is the key to a practical, sustainable agriculture if mankind is ever to adopt one. For forty years we have tried to apply assembly-line effficiencies to farming, coached by simplistic assembly-line economics. These efficiencies, it now seems apparent, don't work in farming; they don't even work very well in factories. But to argue that point any longer appears fruitless. A more hopeful course would be to bring civilization's attention to bear on this concept of biological efficiency and find out how it might be used to preserve human culture. The starting point is an intense investigation of what traditional farming has learned by trial and error over centuries of experience, even if that means humbling ourselves to the contemplation of cowpies.

Traditional farms have several characteristics by which they are known, but above them all hovers a general characteristic in which all traditional practices find their rationale: the traditional farm can survive crisis. It can even survive a series of crises. The urban populations of the Scandinavian countries would have starved to death in World War I and again in World War II were it not for the fabric of Scandinavian rural life; their small farms could go on producing at least sufficient food for the populace even during war. That is why the Scandinavians heavily subsidize their small self-subsistent farms and actually use economic sanctions to penalize their large factory farms. These people know which side their bread is buttered on. Literally.

Traditional farms exist in this country, too, and more farms make use of at least some traditional practices. But the *partially* traditional farms suffer from a principle basic to any biological system: leave out one strand of the fabric of traditional farming and the whole system falters. One either recognizes all the intertwining webs around the cowpie or one drifts gradually into assembly-line economics. This is the wisdom of Amish philosophy. It may be impossible, for example, to have a sustainable agriculture that does not rotate crops in some way; it seems that whenever monoculture violates the rule of rotation, agronomic trouble follows. And while bad weather is going to strike sooner or later, a cash-grain farm is more vulnerable to weather problems than a mixed livestock and grain farm. The greater the variety of crops and livestock, the less vulnerable the farm; the more specialized, the greater the vulnerability. Pasture and perennial crops are less vulnerable than cultivated crops; hail, for instance, does not ruin pasture. As insurance, the traditional farmer has always kept enough extra feed in the barn to last through a bad year. A traditional saying sums up the principle aptly, but modern farmers, following assembly-line procedures, have thought they could ignore it. "Don't put all your eggs in one basket."

Traditional farmers keep their eggs out of one basket, starting with the way they finance the purchase of land. Using the old county biographical histories, I once made a study of first- and second-generation farmers in various midwestern communities. In nearly every case, beginning farmers have had to generate cash from some occupation other than actual farming to pay for their land or to get through lean years. Even where it seemed that the farm was being paid for by farming alone, the farmer was accomplished in some specialized skill that appeared to be part of his farming because he did the work at home. He might have been a sawyer. Or he ran the threshing machine for the neighbors, or operated a seed cleaner. The farming alone did not pay for the farm. Since the industrial revolution, at least, farming has had to operate in an economy geared to manufacturing, with money growth (interest) tied to factory production capabilities, not to rates of biological growth. Science can push a cow's milk production up higher (always at the sacrifice of some other biological attribute) but never as fast as the accelerating exponential interest rates of money, especially when under inflationary pressures.

But the traditional farmer is hardly ever aware that there is a difference between money growth and biological growth. As with all his practices, he either instinctively understands, or is taught by a cautious father, that to be crisis-proof he mustn't borrow large sums of money. The most dramatic difference between the traditional farmer and the assembly-line farmer is that the latter, for better or worse, has had no qualms about taking risks with borrowed money. And by 1982, he was in trouble. Again, traditional farmers have a saying to cover the situation, a saying too often thought of as only a droll bit of folklore from Shakespeare: "Neither a borrower nor a lender be."

Because he won't borrow big money, the traditional farmer is content with a small farm, smaller at least than the "factories in the field." He prefers a small farm anyway, choosing to use biological energy in place of machine energy whenever practical, just as all craftspeople do. Biological energy is limited in quantity and in quality, whether one is making milk or making furniture. But the actual size of the individual traditional farm can therefore vary for the same reason: one farmer, like one cabinetmaker, will have more skill, more energy, more desire than another. I once asked an Amish farmer who had only twenty-six acres why he didn't acquire a bit more land. He considered his wife heaping the table with good food, his sons hoeing the corn with him, his ten fine cows, his springwater running continuously by gravity through house and barn, his few fat hogs, his sturdy buildings, his fine flock of hens, his plot of tobacco and acre of strawberries, his handmade hickory chairs (which he sold for all the extra cash he really needed), and he said, "Well, I'm just not smart enough to farm any more than this *well.*" I have a hunch no one could.

A second characteristic of the traditional farmer, implicit in the one just mentioned, is his or her acquisition of many farm-related skills. He can build barns and houses and knows how to grow the wood to build them with. He is a fair veterinarian, an expert mechanic and welder, can wire, paint, and plumb a house, pour concrete, ditch a field, butcher a hog, and fix almost anything with baling wire and a pair of pliers. My neighbor Bob Frey built his own tractor with tires from a junker, two transmissions from wrecked cars, a motor from a burned-out truck, and a frame he welded together himself. It has twenty-two forward gears and has been running for twenty-seven years.

A third characteristic, implicit in the second, is that the traditional farm is a place of varied enterprises. To spread his labor effectively over the whole year, the farmer has many, but relatively small, sources of income rather than one or two large sources. Elmer Lapp, who farms in Pennsylvania, sells horses, cows, milk, hogs, honey, eggs, guineas, pigeons, chickens, fruit, ice cream, flowers, collie puppies, cats, and tours of his farm. There are even fish in his horses' water tank. With this variety of enterprises, Lapp's work involves a marvelous synergy. When he is hauling manure, he is not only fertilizing his field and cleaning out the barn but also saving the cost of purchased fertilizer, adding organic matter to the soil, controlling erosion, exercising the horses, training a new colt, making the cows comfortable and their milk purer, and keeping his eye on the life of his farm. I have noticed that farmers who do not have manure to haul in slack times tend to spend the hours in cafés complaining about how poor farming is these days. With varied enterprises, the traditional farmer has something to sell every month. Tradition long ago solved for him a concept the computer boys think they discovered: cash flow.

But none of the enterprises is so large that extra money must be spent keeping track of them. The traditional farmer does not need to employ an accountant. He rarely must go outside the family to hire help. His herd of cows is not so large that he needs a computer to figure his feed rations or an electronic eye sensor in the feed bunk to dole the rations out properly. One dairyman I recently visited told me he didn't even believe in the expense of keeping Dairy Herd Improvement Association records. "I *know* which of my thirty cows is giving enough milk and which ones ain't," he said. Nor does the traditional farmer ever have to put big money into new, large models of machinery. Smaller, older ones will usually work in his small fields. For many jobs, horses are even better. Because he deals in smaller amounts of many different things, the traditional farmer usually has time to let his grain dry naturally, avoiding the considerable expense of artificial drying. Field-dried small grains and crib-dried ears of corn are generally of better quality and eliminate the risk to the buyer of receiving grain overheated by an elevator drier or ruined in a moldy pocket in an elevator tower; commercial feeds also are often full of weed seeds. Because he feeds his grain mostly to his animals, he has no large trucking costs, nor is he paying the elevator's storage fee or handling fee for purchased grain. He avoids expensive supplements that supposedly speed the fattening process or the milk flow. He has found his own good grain and hay to be sufficient. He has only so many hogs, sheep, and cattle going to market every year anyway, and it is not important that they all fatten up quickly and are ready to sell at the same time. He has learned that selling a portion every month works out to an average price generally as good as that gotten trying to hit the high markets some expert wants to tell him about—for a fee.

In the field, he is just as financially conservative. He follows a traditional rotation alternating row crops with hay and pasture crops, applying manure on the old hay sod ahead of corn, following corn with soybeans or small grains like wheat, oats, or barley, and interplanting clovers and grasses in these small grains to come on for hay and pasture the next year. These rotations enable him to spend about half of what the assembly-line farmer spends on fertilizer and herbicide; they also solve many insect, disease, and soil-compaction problems.

My neighbor with the homemade tractor follows the old economy of grazing animals all over the farm at appropriate times, so that in addition to all the other synergistic activities taking place in rotated crops, the grain crops themselves also double as temporary pastures. He takes the sheep off the permanent pasture in July and puts them in the wheat stubble after wheat harvest, where they nibble the new clover growth, clean up weeds, clean out the fencerows, and glean any grain left by the combine. Then he moves them to the oats stubble right after harvest, where they do the same as in the wheat stubble. The fattening lambs go into the standing green corn in August, where they eat the lower corn leaves (making corn harvest an easier job), eat weeds grown up since the last cultivation, and clean out the fencerows. They reach up and eat only a few ears—the only grain they will eat before going to market. Later these lambs go onto the old hayfield that will be plowed for corn and will graze there until November. In the meantime, the old ewes have gone back to the permanent pasture and then, after corn harvest, into the cornfield to clean up ears the picker missed and to winter over on the stalks until January. All the while, the animals are dropping manure and preventing heavy worm infestations by moving so frequently onto new ground. What is the cumulative biological efficiency reached by such a synergistic pasture system? Neither economists nor agronomists have recently investigated this practice or how it might be improved.

Traditionally, cornfields have been harvested entirely by animals. The lambs eat the lower leaves, then the hogs are turned in to "hog off" the ears, after which the beef cow mothers clean up the ears that remain and winter over on the stalks. Amish farmers winter horses on stalks, too. Although hogging off may involve some waste, it would be interesting to know how animal harvesting compares economically to machine harvesting, hauling, and drying today. Contrary to what is usually thought, the earlier (traditional) editions of F. B. Morrison's classic *Feeds and Feeding* (1936) report trials indicating that hogs during their first four months fatten just as well on ear corn as on milled corn.

Some traditional farmers still grow open-pollinated corn instead of hybrid, claiming that they can fatten the same number of livestock with less of it than of hybrid corn. Therefore, they are content with the yields of ninety to one hundred bushels per acre that they get from planting at cost-saving low populations—15,000 to 17,000 kernels per acre rather than the 26,000 of commercial, assembly-line farmers.

At any rate, saving seed is a traditional way of saving money. With certified hybrid seed corn over fifty dollars a bushel, saving open-pollinated seed and selecting a strain from it that has adapted to one's soil over the years represents considerable savings—and perhaps a certain amount of security. When the race-T strain of southern corn leaf blight swept through the cornfields in 1970, virtually all the hybrids, because they contained Texas male sterile cytoplasm, were susceptible; open-pollinated corn, however, was unaffected.

Certified or cereal grain seed is not necessarily better, despite the insistence of the sellers. In 1982, my open-pollinated corn yielded more than my hybrid right beside it. A farmer down the road told me that oats he grew from his own seed outyielded the certified seed of a new variety he tried this year. John Vogelsburg, in Kansas, harvests his own alfalfa seed from an adapted variety grown on his farm for thirty-five years and says he sees no reason to change yet. Some of the new varieties of red clover seem much improved, but new or old, this seed can be harvested in humid areas such as Ohio, where alfalfa will not mature properly. Red clover will occasionally yield four bushels per acre, but even a good yield of two bushels per acre means another hundred-dollars-per-acre income from a field that, because of rotation, has already produced a hay crop—another example of biological efficiency long recognized by tradition.

Traditional barns are a marvel to study from the standpoint of biological efficiency. These barns are looked upon simply as objects of beauty today (how does the computer program for that kind of profit?), but they are also extremely functional; their design has evolved through centuries of experience. The upper regions hold feed; the lower floor, often below ground on the north or west side, houses the animals. With their body heat, plus a built-up manure pack of bedding where appropriate, their quarters are comfortably warm in winter without expensive fossil-fuel or solar heating devices. In many of the old barns, as far north as Wisconsin, water tanks are insulated with sawdust above the livestock quarters. On windy days, windmills pumped the tanks full of water, from which it flowed by gravity wherever it was needed. Pennsylvania farmer Elmer Lapp's well pump is run by a waterwheel in the creek next to the barn.

Animal power can lift the hay into the mows of traditional barns, from which, by a thoughtful arrangement of chutes, manpower can easily fork or shovel it back down daily to the animals. What would a detailed comparison of energy use in a traditional barn versus a "push-button" confinement barn reveal? Suffice it to say here that the traditional barn was designed to make the care of all kinds of livestock a practical and efficient chore for human labor. A confinement operation, on the other hand, is designed to handle many animals of just one kind, requiring sophisticated machines and power sources, vulnerable to breakdowns and failures. When a poultry confinement building in Oklahoma lost its electricity in a storm last summer, causing the shutdown of a ventilating system, thousands of chickens died in less than fifteen minutes. Contagious disease is another major vulnerability of confinement feeding operations; prevention of disease costs the farmer another heavy expense in drugs, the long-term effects of which are unknown. But the greatest vulnerability of the specialized confinement feeding system is that it locks the farmer into an investment in buildings and equipment that can be used only one way. In 1981, farmers were losing ten dollars and more on every hog they sold, but were forced to keep their assembly line moving because of debt commitments. This is the inane tyranny of applying factory economics too literally to farming.

The draft horse provides a good example of the logic of traditional versus assembly line economics. By the criteria of assembly line economics, work horses seem to cost more than tractors, and the arguments pro and con have continued for fifty years. True Morse, then president of Doane Agricultural Service, Inc., made one of the first explicit statements of the assembly-line viewpoint in 1946 in *Farm Journal* magazine. He argued that although it did cost more in actual cash outlay to keep a tractor than a team of horses (even then), the farmer would be ahead if he sold his horses and bought two cows in their place. Farmers took that advice, sure enough, and we have had a surplus of milk ever since, along with an army of bureaucrats to administer a tax-supported market subsidy program whose agenda, USDA economists have admitted to me, is "too complicated to explain." But that point aside, the divergence of view between biological efficiency and factory economics as it pertains to horses stems from how each side defines cost. Morse put the annual cost of keeping a team in 1946 at $182.49. (The traditional mind immediately suspects such a finely wrought figure. Why not $182.50? A computer might be forgiven such preciousness, but not anyone who has smelled horseflesh. I find myself wondering if Mr. Morse figured in the wear and tear on the harness, or the fact that old Bell hangs back in the traces worse than Flora; or if he counted the penny of work the dung beetles contributed to hauling off the manure.) But even in the unlikelihood that horses can be summed up by numbers so precisely, what does this cost mean? To a farmer who loves horses, a good team gives a satisfaction akin to what the modern playboy gets from his beloved Corvette. It is a cost that might easily be reckoned as a profit, for horses are worth much more to the horse lover than the money he might make milking two more cows the rest of his life.

More importantly, most of Mr. Morse's cost is not out-of-pocket cost, a distinction an accountant is bound to scoff at but not a farmer who knows, especially in the 1980s, what it means not to have any cash in his pocket. When that happens to a tractor owner, he still must find money to pay the tractor bills. He has to borrow it—if he is "creditworthy," as the bankers say so unctuously. (If he is creditworthy, he has the honor of paying interest—which has reached 15 percent in recent times.) A horse doesn't charge a farmer interest. The bartering system between the horse and farmer is immune to inflation and depression. The horse trades its labor, manure, and affection for food and shelter, then throws in a free colt every year or so that is generally worth more than any of the farmer's calves. No cash outlay is involved at all except for veterinary bills. The horse will start every morning no matter how cold the weather, and it will run all day no matter what happens in the oil fields of Saudi Arabia. That is the kind of security that biological efficiency provides. Beats bankruptcy any day.

Mr. Morse perpetrates another piece of assembly-line logic. The cost of the horses per hour, in 1946, was determined to be about thirteen cents; the cost of the tractor per hour was about forty-seven cents, over three times greater. But, Mr. Morse reminds us, the tractor did the work of eight horses (that's arguable—the 1946 tractors I've driven will hardly pull a loaded manure spreader in winter snow and are almost useless for dragging logs out of the woods). Assuming Mr. Morse is correct, he nevertheless fails to mention that the tractor also took the place of three or four farmers. When farmers chose tractors, 75 percent of them chose their own demise, their land becoming someone else's factory in the field farmed at three times the cash outlay they had previously spent farming it.

How much money would farmers have saved by not buying tractors between 1946 and 1982? Tractors rose in price from $800 to $18,000 for comparable models. The horses kept having colts. So, twenty horse farmers, each with one hundred acres, would have had no cash outlays for new models. The modern tractor farmer, going broke on the 2,000 acres those twenty farmers would be farming, owns (with the bank) around $200,000 worth of tractors. In 1955 another farmer, Raymond Rall, leaned over his tractor fender and said to me bitterly, "Farming with tractors, I'm spending up all the money I saved farming with horses." Ironically, yet another neighbor, Jerome Frey, who has kept his horses just for fun, confessed the other day that he believed he could make just as much money farming eighty acres with horses as on five times that much land with tractors. "But, heck," he shrugged, "I'm kind of stuck with all that land now."

The net profit from the two cows to replace the team of horses in 1946 has not kept pace with the surging tractor prices, either. Mr. Morse quotes a cash return per cow in 1946 of $119.50 per year. According to *Dairyman's Digest,* a cow in 1982 giving 16,000 pounds of milk with a butterfat content of 3.5 percent netted $106. This figure deducts $318 per cow for labor, a deduction Mr. Morse probably did not cipher into his cash return in 1946, but nevertheless, cows have hardly proved to be that much better an investment than horses. Using assembly-line economics, one must now keep a hundred cows, if only to pay the labor it takes to run the assembly line.

The traditional mentality approaches profits differently, substituting biological quality for assembly-line quantity. For ten years, Rex and Glenn Spray in Ohio have bred their Holsteins for high butterfat content, not high production—in direct opposition to the advice of assembly-line economists. Their thirty-five cows today average 4.1 percent butterfat. With a premium of 17.1 cents per point for every decimal point above 3.5, their six extra points amount to an extra dollar per hundredweight for their milk, or an extra $160 from a cow giving 16,000 pounds of 4.1 percent milk. On thirty-five cows, that is a $5,600 annual premium at no out-of-pocket cost at all. With the current figure of $106 net profit per cow, to make that $5,600 from production alone requires another fifty cows.

Once the minds of farmers and scientists open to the possibilities inherent in biological efficiency, what wonders will follow? In an example so outlandish one suspects it is a joke, *Audubon* magazine reports that a greenhouse grower in Oregon is heating his greenhouse with rabbits. I assume he puts the manure to good use, too. Experiments at the Rodale Research Farm in Pennsylvania and at the University of Southern Illinois indicate the feasibility of an old Oriental tradition: fish can be fattened directly on the manure of chickens and hogs, and the pond water then makes an excellent fertilizer. The Chinese get amazing protein production combining fish and ducks in the same pond-rearing system.

As traditional American farmers pursue the "let nature do it" methodology, they may find, as I have, that they edge closer each year to a type of farming based upon permanent or semipermanent pasture arrangements and lesser amounts of annually cultivated land. The key to general success in this arrangement is a change in marketing standards. Like myself, the Kansas farmer Oren Long raises a Jersey-Angus cross of baby beef (I have a Guernsey-Angus cross) that when butchered at 700 pounds makes a most delicious meat, fattened on ample milk and pasture but no grain, the mothers getting only a little grain during lactation, or none when pastures are lush. (In addition, we get our own milk and cream from the cows while the calves are fattening, for greater biological efficiency.) But this meat will not be graded as "choice" because it is not grain-fed. Therefore, the farmer cannot sell it profitably in the conventional market. He must develop his own customers who know good meat when they taste it. In the same way, our lamb is luscious even to people who previously said they "hated mutton," and we have produced it quite successfully on mother's milk and weedy rough land—without even good pasture, let alone grain.

Small-bred cows, like Jerseys and Guernseys, will produce a fair to good amount of high-quality milk on good legume hay and very little grain. Heavens, if the public wants low-fat milk, as the milk marketers say, add water to it after it leaves the farm. Why make the cow do it? Then those who appreciate good rich cream and creamy milk can have their desires met in the marketplace, too. More and more of the milk, and especially the cream, in stores now is "ultra-high temperature" pasteurized stuff that will keep without refrigeration but bears a bitter aftertaste (not to be confused with the off-taste of milk kept in plastic jugs). The dairy industry wonders why milk consumption is down. The crowning irony is that I cannot sell my own milk from cows checked annually for tuberculosis and brucellosis as in any Grade A dairy, cows who have never had mastitis and never been treated with antibiotics. It is against the law in Ohio to sell raw milk, no matter how good and healthful it is. The law says I can drink it, but my neighbors can't, which reveals the true purpose of the law: to stop me from underselling the assembly-line milk production system, which I can do because of the biological efficiency of a very small operation.

Even nongrazing animals make excellent use of semipermanent pastures. Hogs can fatten profitably, says Morrison, with up to one-third of their diet derived from alfalfa pasture. A small flock of hens can find half its food or more ranging the barnyard and adjacent fields. I have halved the amount of grain even broilers need, ranging them outside this way. The only difference is that they take a little longer to reach butchering weight.

If marketing standards were changed, a successful farm could be maintained at a ratio of five acres of semipermanent and permanent pastures to every one acre of annually cultivated crops, the whole not exceeding about 200 acres and probably working best with about 120 acres, of which twenty were woodland for lumber and fuelwood. At this size, biological energy could be substituted profitably for the large inputs of mechanical energy necessary on the assembly-line farm. A great variety of products could be sold: milk, cream, lambs, wool, beef, eggs, fuelwood, lumber, honey, fruit, broilers, hogs, surplus horses, old cows, and hens. Of the 120 acres, only twenty would be in annual cultivation, and that twenty would be divided into three fields rotated from corn to wheat or oats to legume hay. The cultivation of these three small fields could easily be handled with horse power or very small tractors. The other one hundred acres in permanent and semipermanent pastures would need to be fenced into relatively small plots, the animals moved from one to another regularly after the New Zealand method. Hay would be harvested from about half the pasture whenever the grass and legumes grew faster than the animals could graze them. Haying would be the farm's heaviest labor and fencing its highest cost. Vogelsburg's haystacks and New Zealand–type electric fences are two answers to lessening these burdens. More will come.

But providing the technical information from tradition that would help forge a modern sustainable agriculture is not the most difficult task. Much of the technology is there waiting to be reimplemented. What we lack most is the traditional farmer and a culture that could nourish him.

Before the industrial revolution, corn shocks were hauled in good weather to the barn, and then in harsh winter, the young people went from farm to farm in the evenings making a party out of the husking. The person who husked a red ear—and there were many red ears in the days before standardized hybrid corn—got to kiss his or her sweetheart. This was the cultural, even cultured, way of making work pleasant. It was replaced by a farmer husking corn alone in a cold December field, day after day—a misery, one he was driven to when technology made communal work impossible and obsolete, and when traditional social rituals had lost their significance. Humans are social animals, and it is as community, not as individuals, that they will accept a life that eschews the consumerism, leisure, and delirious pursuit of novelty that characterize our society today. Only as a community will people rediscover the unique kinds of pleasure and satisfaction that the traditional farm provides.

The reason the so-called back-to-the-land movement failed for so many people is not so much that they were ignorant of traditional technology—too stupid to stuff rags in a rat hole, as country people say—but that their homesteads were islands in an alien culture. There was no community to rebuild their barns or their dreams. Indeed, their retired parents, not understanding that the culture they had forged had lost its meaningfulness for younger people, were often hostile to such "outlandish" dreams of forsaking the American way of luxurious consumption. Then the homesteaders' own children, out on the land, rebelled against the physical labor of homesteading, not because it was hard work but because there was no cultural reward or recognition for doing it. And that made all the difference. If we want to remake an agriculture that is technically correct for sustainability, we must make sure it is also culturally correct, or the effort will not succeed.

6

Knowing One's Place

1991

Dave Haferd sees his farm with eyes that are 200 years old. He knows every foot of its 180 acres, on top and underneath. Walking across his land, he discourses endlessly and joyfully upon almost any rock, post, tree, clod, weed, or building that his eye falls upon. The gully that cuts deeply into the hill going down to the creek is where the road used to go years and years past, he says. The boulder in the fence corner required two days of hard work to move out of the field, he says, which reminds him that over in another field—he waves his arm in a southerly direction—there is a stone so huge imbedded in the soil that he has never been able to move it. He worries, now that he is thinking of retiring, that the next farmer will break his plow on it.

The wild hop vines on the fence are not really wild, he confides, but escaped years ago from the fields, when hops were grown here commercially. He says this casually, not seeming to realize that he may well be the only person, until now, who possesses this potentially useful memory of this northern Ohio county.

Over there, across the boundary of his land, on what is known as the High Bank of the Tymochtee Creek, he says that the Indians burned Colonel William Crawford at the stake in 1782. "Or so the history books say," he adds. "Actually, I believe Crawford was burnt in the bottomland across the creek from the High Bank. That's what Black Betty told my grandfather. She was a herbalist who often came to the farm in the late 1800s. She told grandfather that she had talked to Indians who had been there."

The boulders set at regular intervals in a loose line across the Tymochtee, he points out, were put there for stepping-stones by the Indians and were, he believes, part of the ancient Indian trail known to have traversed this region. And just down from the stepping-stones is the old ford, where, before good bridges, farmers drove their horses and wagons across the creek.

Walking along the edge of one of his fields, he asks me if I can see anything unusual in the wheat growing there. I cannot. "If you look close, you can see a sort of division. On the west side, the wheat is a bit taller and lusher than on the east." Now that he points it out to me I can see the difference. "On the east side," he explains, "the land was cleared and farmed eighty years ago, and on the west side, forty years ago. I still call the west side the 'new ground.'"

In an isolated little cemetery we walk through, he pauses at almost every tombstone to give a brief history of the grave's occupant. "That fellow was worthless," he said. "And that one next to him hit Poppa with a hoe handle over a line-fence dispute."

In another field, he stops suddenly and studies the ground. "Right here someplace there's an old gas well. Pipe broke off down in the ground but the gas continued to seep up to the surface for years. We would light it when we were hunting and have us a real nice campfire."

It is not only old bones and gas wells that he knows about. "There's an eight-inch tile runs through under the fence right there and goes clear across that bottom ground to the hill, with four-inch laterals branching both ways along the foot of the hill," he says, as if I were the son he never had, the next generation to whom this essential knowledge needs to be passed on. "Well, I've got 'em all drawn out on a map," he says, almost to himself, "but it isn't the same as coming out here and seeing where they are." He pauses. "You really can see them some days, you know. Right after a rain, on cultivated soil, the dirt will dry out first right over the tile lines."

Strolling along the creek that flows through his farm he speculates on whether, as rumor has it, the landfill upstream might be polluting the water. Although the idea troubles him, he is not given to the shrill protest that abstract knowledge brings to environmental debate. "There's still lots of fish in the creek, so it can't be too bad," he notes hopefully. Then he smiles and the patience of a thousand years of peasantry glints in his eyes. "When I was a boy this creek ran black with oil during the oil boom years. The oil scum got blocked by fallen logs and my brother and I set fire to it. The whole creek was on fire. A sight to behold." He pauses to enjoy the scene in memory one more time. "But forty years later we drank out of the creek again."

Today, with our wives, Dave and I are returning from a walk

viewing his crops, our Sunday afternoon ritual. As we pause on the road in front of the house, he points with his walking stick at a barely discernible, grassed-over rut that starts at the foot of the hill next to the cornfield below us, and runs parallel to the road up past the grove of pine trees, then along the edge of the garden, and disappears where it meets the lawn at the top of the hill. "That's the old cowpath," he says. I nod, although I can scarcely make it out. "That's where we used to drive the cows up from the creek every evening. Right up the hill across the edge of the garden and lawn to the driveway, then on to the barn. After the cows learned to follow the route, we didn't even have to worry about them scattering off into the yard."

We walk on to the driveway then, myself marveling at the detailed knowledge Dave has of the farm he was born on, had farmed all his sixty-four years, and which his father had been born on and had farmed his entire life, learning from *his* father, who had spent the greater part of his life on the farm, too. "See that bit of depression in the grass right there?" Dave asks. "Used to be a hitching post there. Sunday visitors would tie up there and the horses wore a hole in the ground stomping their feet to chase away flies."

I stare at him in great wonderment. Although we have been close for many years both by reason of kinship and friendship, or perhaps because of that, I have never been able to convey to him the uniqueness and significance I see in the depth of his knowledge about his farm. It is something he takes for granted, as if everyone knows their places as well as he knows his. I do not know how to tell him that he is a last member of an ancient tribe—the genuine traditional farmers who committed themselves lovingly to a piece of land and husbanded it from generation to generation, carrying in their memories a lifetime of their own experiences and that of their fathers and grandfathers on that land. Dave's crops are almost always just a little better than the others in the neighborhood, because he knows his place.

Had I spoken all that aloud, Dave would only have been amused at this example of what he takes to be my romantic exaggeration at work. He does not know his own value. He does not know that the disappearance of his kind puts society at terrible risk. A stable food supply depends entirely on the concrete particularity of his kind of knowledge, without which the abstract expertise of science is useless. If one can leap imaginatively over the ruins of several civilizatons, there is a direct line between Dave Haferd and the ancient Phoenician settlers who turned the sands of North Africa into a garden of plenty, building precisely on this same intimacy with place. When those farmers disappeared, so did North Africa's glory days, leaving not only an agriculture in ruins, it needs to be pointed out, but a landscape dotted with empty amphitheaters. (Read: football stadiums.)

But after walking with Dave Haferd on his land, there is comfort even in that. If errors are repeated over and over again, rightness returns again and again also.

We wade into Dave's clover field west of the barn. It stands almost to my thighs, hardly a weed in it, our noses full of the sweet smell of its blossoms, the air over it full of dancing butterflies and bees. Redwings and meadowlarks rise from it and settle back into it.

Red clover is always in the rotation on this farm, whether Dave makes hay from it or not, and in these latter years more often he just plows it under. "Oh yes, it still pays if you only plow it under," he says. "*This* ground would not last without a regular plowdown of clover. And it's the greatest help in weed control. We learned that long ago."

The clover reminds him of something that makes him smile. "We always sowed clover seed by hand, walking. I still do, in fact, most of the time. Poppa was not one to buy new machinery if he could avoid it. We had one of those little hand-cranked seeders that hangs on your shoulder, and a fiddle seeder, and one of those horn-type broadcasters, which I guess was the oldest of all. He'd assign the crank model to me, the horn seeder to my brother, and he'd manage the fiddle seeder. Side by side we'd walk back and forth across the field, sowing seed, cranking and fiddling and sawing the air. From the road it must surely have looked like a three-piece band marching along."

7

The Future: More Farmers, Not Fewer

1989

Although the percentage of Americans engaged in farming has been decreasing steadily since the ink was hardly dry on the signing of the Louisiana Purchase (from 90 percent of the population then to 2.5 percent today) and has been decreasing in actual numbers since 1916, the one continuous viewpoint I have heard in my fifty years among farmers is that the number of farmers can't go any lower. I have personally known all the editors of the prestigious *Farm Journal* except the first, who started the magazine 110 years ago, and at some point in their careers, every one of these astute gentlemen opined editorially or privately that the decline of the farm population was about to bottom out. Even Wheeler McMillen, who in the thirties wrote a book titled *Too Many Farmers* (thereby earning himself the undying enmity of small farmers everywhere), thought that the right number to get down to was about 9 million, more than four times what we have today. And of the 2.2 million remaining (or whatever the number is this week), only about 700,000 contribute significantly to the commercial food market. All of which at least means I am in good company when I continue the naïve tradition of optimism in the face of statistical and economic reality and declare without batting an eye that the family farm is not dead, and that, far from continuing to decrease, farms and farmers are on the increase.

Although megafarms will continue to be an important part of the food production business, I don't think the future belongs to them at all, despite the predictions of wishful-thinking agribusiness interests. The real action is going to occur in the comparatively small-scale food production systems now sprouting up everywhere, and in those which have handily survived the economic crunch of the eighties. In short, there is no better time than right now for dedicated young people, determined to own and operate their own businesses, to make it in agriculture.

I can give three reasons for my prediction that the number of food and fiber producers ("food and fiber producer" is the only definition of farmer that works) is about to increase. One: historically, in all the past civilizations I have studied, the denser the population becomes, the smaller and more numerous the farms become. Two: financially, the economies of scale that apparently rule manufacturing do not really apply to any sustainable kind of food production; when you count all the costs, it is cheaper to raise a zucchini in your garden than on your megafarm. And three: socially, people are beginning to understand they really are what they eat and are demanding quality food, which megafarms can't supply.

New attitudes toward food are not only increasing the number of farms but, more importantly, are bringing a new kind of farmer to the land—a farmer with roots in urban culture, not traditional rural culture. While commercial agriculture stolidly continues to pile up government-subsidized mountains of surplus hybrid corn and hard red wheat, these new farmer-entrepreneurs bend an ear to the marketplace and produce the food consumers want. They are often called specialty farmers, but they are more apt to refer to themselves as "guerrilla marketers"—they strike where the big boys aren't looking, and they come in a variety that is mind-boggling. The *New York Times* food writer, Marian Burros, noting that "there's not a New Englander or farmer in the lot," recently listed such disparate new farm products in New England as goat cheese, farm-raised oysters, hand-pressed cider, hydroponic spinach, stone-ground flours and meals from locally grown organic grains, baby lambs for the hoity-toity restaurants of New York and Boston, pheasant and other once-wild game such as buffalo, and dairy sheep producing Roquefort cheese.

All across America, hundreds of small sheep flocks have come into existence in the last ten years to supply wool directly to the bustling cottage hand-spinning and weaving industry. Angora goat farms have become almost common, even in the corn-and-soybean wastelands of the Midwest. A feedlot in Nebraska now raises ostriches, some of which bring $22,000 a pair. In Michigan, Juliet Sprouse told me last year that female llamas she and her husband raise on their thirty-five acres had skyrocketed in price from $1,500 six years ago to $10,000 now. New Jersey vegetable farmers, taking a cue from the West Coast, are learning to market squash flowers for food as well as squash fruit. Catfish farms thrive in the South as consumers learn how good a humble, down-home fish can taste when raised in unpolluted water. Crayfish farms are on the upswing in the South, too, and very likely, if humans persist in using their rivers and oceans as a sewage-disposal system, *all* fish and seafood will eventually be raised on domesticated "farms."

Samuel and Louise Kayman are good examples of specialty farmers developing new markets. In 1983 they began making high-quality yogurt from their herd of Jersey cows at Stonyfield Farm near Wilton, New Hampshire. (Sam had quit his job in the defense industry in the sixties.) Their market exploded and they could not keep up with demand. The Kaymans sold their cows and concentrated on the yogurt; they bought Jersey milk from surrounding farmers, even paying them a premium over the outrageous price the tax-funded dinosaur government-subsidy program allows farmers in the conventional market. They brought in Gary Hirshberg from New Alchemy Institute to provide financial and overall management. By 1988 they had hit $3 million in sales and moved into a new half-million-dollar plant away from the farm. "Significant profit is still in the future" says Hirshberg, "but there's a tremendous nonmonetary reward we gain from customer satisfaction. We get hundreds of letters from people who love our yogurt, and that's very important to us. We also provide a better profit picture for the farmers who sell milk to us, which is part of our overall goal of revitalizing rural areas. When we succeed at those goals, there's an intellectual payoff more valuable than money."

Those words are a fairly accurate modern translation of the traditional farmer's reverence for and stewardship of the soil entrusted to him. I heard the same echoes from Jay North, who, with his wife, Pamela (and now forty employees), grows about a million dollars' worth of edible flowers and herbs a year—not bad for a thirty-seven-acre farm, even in California. Admitting he didn't have a computer yet, Jay explained: "If I computerized this business I probably wouldn't need as many employees and I wouldn't like that. The success of this business depends a lot on a sensitive relationship between humans and the land. We would rather employ people than machines." North's background? Previously he was a hairdresser.

Thus a historic shift takes place. The Vergilian ecology and careful husbandry of the traditional yeoman farmer that gave way to the all-consuming dreadnought of agribusiness economics now reappears in the unlikely form of the ex-urban farmer. Fleeing industrialism, this new farmer is characterized, writes Marian Burros in the *New York Times,* "by an environmental awareness often influenced by the 1960s counterculture movement: a desire to be free of the constraints of the crowded city and nine-to-five jobs and a belief that success can be achieved by working fourteen-to sixteen-hour days." Jefferson's agrarian dream returns in modern guise.

So many specialty farms are starting up that it is hard to keep track of the number of new growers. Jay North complains that there are almost too many, driving the price of baby vegetables, salad greens, edible flowers, and other high-value crops down to where he fears the profit will go out of them (the old Wheeler McMillen syndrome). He guesses the number in his area has gone from about four to thirty in the last couple of years.

California's *Farmer-to-Consumer Directory* lists about a thousand certified roadside farm marketers. The foods offered include everything from baby beans to nasturtium blossoms to feijoa (pineapple guava). The *Directory* also lists one hundred farmers' markets and twenty-two regional "Farm Trail" organizations that promote their members' locations and products. The story is the same in

New England. While nationwide more than a million farms were lost in the last twenty years, and while more than a million people have left the land since 1980, Massachusetts, of all places, gained over 500 farms since 1978. Maine, despite a very bleak picture in its commercial potato fields, gained over 400. In the Midwest, Minnesota gained 2,000 farms in the latest census and Wisconsin 1,000. Georgia and Florida gained 1,000 each, too.

However, I discount census numbers, up or down. By census definition, a farm is an establishment from which $1,000 or more of agricultural products are sold during the year. There are thousands of homesteads where nothing is sold, but which produce over $1,000 worth of food, including thousands of the more than thirty million gardens in this country. In 1983, some imaginative statistician figured out that the total value of food raised in American gardens exceeded the profit in all of commercial agriculture that year.

It is my contention that gardens are the incubators of the new farm ecology. In Berkeley, California, a city "farm" of one-third acre grosses more than $300,000 in salad vegetables for Konakai Farms. An article in the Salem, Oregon, *Capitol Press* of August 19, 1988, tells about Neil Lawrence's three-quarter-acre "farm" from which he says he makes a living raising high-priced vegetables.

Mark Musick, who now works as a buyer for Larry's Market, a local supermarket chain (and who deserves the title of Guru of the Guerrilla Marketers), works hard as a liaison between the store and garden farmers, some of whom hitherto were too small to interest supermarkets. Just to list a few of the foods he deals in makes the palate quiver in keen anticipation: exotic pepper varieties from Krueger Pepper Gardens in the Yakima Valley; wild blue huckleberries, harvested by Yakima Indian families; yellow finn potatoes and romanesco broccoli from Dungeness Organic Produce; several varieties of red, white, and blue potatoes; Spanish Roja garlic; nettles (along with a recipe for cream-of-nettle soup from Chef Karl Beckley in Seattle); boletus mushrooms and cattail corn from Mike Maki, who has leased 40,000 acres of forestland for commercial foraging of wild foods (making his one of the largest "farm" operations in the world); shungiku and baby bok choy from a co-op of six Laotian refugee farmers called the Indochinese Farm Project. Musick was a longtime farmer at Pragtree Farms, the home of Tilth Publications, north of Seattle, where he helped spearhead the new interest in unique foods, especially salads, as a way to make small farms both ecologically sane and economically profitable. Pragtree's "Impromptu Salad," which contains twenty or more different herbs, salad vegetables, and edible flowers mixed together, costs twenty-eight dollars a pound before packaging, and since being featured in *Harrowsmith* magazine has become nationally renowned. "Before, we were grossing $3,000 an acre on five acres of conventional vegetables. But cabbage at twenty-five cents a pound just didn't pay. What we were able to do by switching to lightweight, high-value specialties like salad vegetables, herbs, and edible flowers was reduce farmland from five to one-half acre, sell both tractors, hand-dig the plots, and still make the same amount of money."

Nor is the new yeomanry confined to the far-out East and West coasts. In New Mexico, Richard Deertrack has made a success of selling traditional blue corn flour by mail—from the crop grown by his fellow Pueblo Indians. In southern Indiana I watched a remarkable friend, Ed Fackler, struggle for years to establish a commercial organic fruit orchard. He had previously worked for the railroad. For anyone aspiring to the ranks of ex-urban yeomanry, he would make a good model, and I can tell you his secret in four words—intelligence plus pure mule-headedness. In time, he mastered everything known about organic orcharding and finally concluded, after eleven years, that he couldn't raise fruit that way in his humid southern-Indiana location. But rather than give up his high environmental ideals, he worked out an orchard agronomy that uses only a small portion of the toxic sprays conventional orchardists regularly douse their trees with. He started a mail-order nursery, emphasizing trees and varieties that perform well in his locale. He also started selling richly flavored old and unusual varieties of apples and other fruits to the Louisville market. The Louisville *Courier Journal* did a story on his unusual apples last year, and he sold out completely in ten days, at premium prices. "People went ape over these different kinds of apples," he says. "We had customers from as far away as Bowling Green, Kentucky, and central Illinois. One broker drove down from Cincinnati to set up next year's purchase of 200 bushels of Scarlet Gala apples and all the Asian pears we could furnish. At the same time, growers nationwide were taking a beating on standard apple varieties, which were in oversupply because, in my opinion, they have poor fruit quality." Fackler says that he began to pay special attention to the Gala apple, now his favorite, when he noticed that his little daughter would invariably pass up the other 500 or so varieties in his experimental orchard to feast under the Gala tree.

Parker Bosley, of Parker's Restaurant in Cleveland, Ohio, [closed because of fire in 1991] has played the same role in Cleveland that other innovative chefs like Wolfgang Puck and Karl Beckley have on the West Coast. He went directly to the farms and told them what he wanted and how he wanted it—and that he was willing to pay for it. He established what *Ohio* magazine calls "an old farmboys' network," reminiscent of Mark Musick's program. Bosely's network supplies him with, among other things, rabbits and ducks from Ashland County and lambs from Portage County. Two women from Cleveland Heights grow blue oyster mushrooms in their garage for him. He buys veal from Geauga County, berries from Lorain County, and thick Guernsey cream from Wayne County. During the growing season he uses all-local produce and pays premium prices for it.

Obviously, people who eat edible flowers or pay twenty-eight dollars a pound for salad greens are not eating just to live. But in a deeper sense, that is perhaps exactly what they are doing, since these foods are almost always raised without pesticides. After years of hearing from conventional agribusiness that consumers would not pay for quality food or (heaven help us) organic food, quite the opposite is suddenly true. This is new. For years the image of the "good" little housekeeper was bargain-hunting Mom, trundling off to the grocery store to take advantage of every coupon, every sale, every loss-leader offered. Virtue as a good shopper consisted of finding the cheapest food in town. Now, more and more, food is seen as an enhancer of health or beauty or status, as having not only intrinsic health value but also recreational value like beer and whiskey. While it is true that the keen edge of demand comes from the social levels that have lots of disposable income, people in general are deciding good food is worth paying a little more for.

Whether "organic" or simply free of pesticide residue, or just plain fresh and flavorful, the specialty crops from the new farms are also deemed by consumers to taste better. "It's not just status-seeking yuppies influencing the market," says Jay North. "Our main customers are women between the ages of forty-five and sixty-five who really know their foods." Gary Hirshberg, at Stonyfield Farm, points out that quality foods are not necessarily so expensive, either. "Not at the retail level. Our yogurt is only about five cents a cup more than the common brands."

At this point, I wish the reader could turn quickly to the first chapter of Jane Jacobs' brilliant 1969 book, *The Economy of Cities.* Jacobs' thesis is that civilization did not necessarily proceed, by orderly evolution, from a hunting-and gathering stage, to domestic farming, to villages, to towns, and, finally, to cities. She argues, instead, that cities actually preceded a genuine agricultural economy. "Cities are not built on a rural economic base. . . .Quite the reverse is true. Rural economies, including agricultural work, are directly built on city economies and city work." Just as alfalfa was grown in French gardens for a century before it became a commercial farm crop, as Jacobs notes, so too we can observe the same process with salad greens, herbs, and edible flowers today. It is cities that have created the market for these new products. Restaurants with innovative chefs inspired the new farmers. Their farms are extensions of the city, willed by the city. And therein lies the reason they must be taken seriously.

The animal rights movement also springs purely from city culture, which is one reason why conventional farmers oppose it so vehemently. Yet if the moderate mainstream of animal rightism prevails—that is, if society still wants to eat meat but insists that animals be treated more humanely and raised more naturally than is now the case in large confinement factories—the result will be good for farmers. It will mean that more of them will be able to derive a decent living from husbandry rather than the condition that now prevails, wherein fewer and fewer megafarms raise more and more of the poultry and pork with methods that depend increasingly on drugs and chemicals. The meat will cost more money perhaps, but it will be worth more, and the new consumer is obviously willing to pay. It is very likely that the huge animal factory complexes will eventually be phased out for no other reason than because, like landfills, nobody wants to live near them. Meanwhile, the state of Vermont has now passed a law whereby traditional family dairy farms will be paid so much per pound of milk produced just to stay in operation. Urban society wants the beauty of the open space that farms provide to remain on the landscape. Even twenty years ago, such an attitude would have seemed highly unlikely.

If my argument for a future of more farms and farmers rested solely on the specialty farms, I would not advance it with so little fear and trepidation. But there is another kind of farming, another kind of guerrilla marketing that, in concert with what I have been describing, makes more permanent something that looks like only a fashionable, momentary trend. To describe this kind of farming, I need only to look out my window across the Upper Sandusky neighborhood I have watched for fifty years and which has always told me the future of farming, when I was paying attention. Next door to me, a couple and their two school-age children recently started milking eight cows on two acres, the sort of spare-time endeavor the universities tell their ag students is not profitable. Fortunately, my neighbor did not go to college and does not know he can't make money milking cows on two acres, so he does it anyway. He wants to be a "real" farmer but fate has not put into his hands the kind of money it takes to own a "real" farm. So he operates a tile drainage business by day and milks his cows by night. He says he comes out in the black most of the time because he has no huge investment in farm land and machinery, and because he can take advantage of the propensity of megafarms to over-produce corn. For eight cows, he can buy feed cheaper than raise it. But as with all real farmers I know, he *really* milks cows because it is something the whole family can do together, something his children can earn wages and learn responsibility from. And, at the bottom line, he just *likes* to do it.

If I swivel my head in a slightly different direction, I see my own barn through the trees, where Betsy and her new calf are presently penned. They represent, among other things, a successful experiment I thought I had pioneered in the production of low-fat, low-cholesterol beef, only to find that my friend Oren Long, a Kansas rancher, had already perfected the method and was selling all the organic baby beef he could raise to supermarkets, who displayed it right alongside the conventional meat showcases—even at three times the price of conventional meat. His beef tests lower in calories and cholesterol than chicken, and he says the taste is unbeatable, a claim I believe, because my own beef, raised similarly, is absolutely luscious. Organic baby beef, as Oren and I produce it, comes from a calf slaughtered at about 700 pounds and seven months of age (conventional beef goes to market at 1,000 to 1,200 pounds and nearly two years old).

These calves grow mostly on mother's milk and good pasture, along with some grain in the last two months, but in my experience, if the calf is getting a lot of rich milk from its mother, it won't eat much grain. No antibiotics. No drugs. No protein supplement. The calves do not get fat, so the meat is not marbled as in prime beef and does not have the heavy, tallowy aftertaste of prime beef.

Consumers have been so responsive to this kind of beef when made aware of it that large feedlots are now trying to reproduce it. It is my contention, and Oren's even more so, that this kind of organic beef (he calls his Lite Beef) *can't* be raised on a large scale. It demands too much personal attention, a steady supply of good pasture, and a mother cow that is a cross between dairy and beef breeds—he uses Jersey-Angus, I use Guernsey-Angus. Says Oren, who's in his sixties, "I'm getting too old to do this much work. There's opportunity here for young small farmers with a flair for salesmanship." The implications are enormous. If beef feeders followed this method, pasture, not corn, would become the chief food in beef production, negating the need for thousands of acres of soil-eroding corn and providing consumers with what they want: low-cholesterol, low-fat red meat.

My neighbor across the road is a plant manager and farms eighty acres. Next to him in one direction is a mechanic who farms eighty acres. In the other direction, yet another part-timer, a carpenter with eighty acres who has just built a new barn. Across the road from him is a retired farmer whose son farms the land as a part-timer. They sell eggs from free-range hens, just as I do. His clientele say they enjoy driving out to his farm and so the cost of transporting the eggs is absorbed by customer satisfaction.

In another nearby neighborhood are five farms almost adjacent to each other that are owned or rented by small farmers very familiar to me. Two are relatively new operations started by men who formerly had other occupations. The wife of one of them has her own small business, which she conducts out of the house. A third farmer carries on a machinery-repair business along with his farming. A fourth recently rented additional land after a mega-farmer got into financial difficulties and had to give the land contract back to the original owner. This renter also has another business. The fifth farmer has always farmed only his family's 200-acre homestead, dividing the income with his brother and sisters, and, contrary to what the farm magazines assert about such arrangements, has done just fine financially, even in these "bad" years of the eighties. The pattern emerging before my eyes tells me that good, conservative, conventional farming is not in trouble at all, because small farmers either avoid indebtedness or have another source of income upon which to draw. Not only will these farms be around as long as their owners live (and then, very likely pass on to offspring), they give every indication of being on the increase. My own brother, for example, using money from another small business to purchase land, has in these so-called hard times realized his dream to become a full-fledged farmer. My niece and her husband recently bought a small farm with their savings. They are in their twenties. Bidding against them were not big farmers, but other young couples like them. In the whole area west of my place, centering around the village of Kirby, there is a long-standing tradition of combining farming with another job or business—I could fill this page with the names of families doing it. Nor is it at all surprising; a study of how farms got started in the Midwest shows that among the pioneering families, farmers invariably depended on cash from another job or business to establish themselves on the land.

What is so intriguing about these small farms is that because many of them carry a relatively small debt load (interest payments are the highest cost per acre on most Midwest farm operations that have expanded greatly in the last twenty years), they can actually raise food cheaper than the megafarms can. The Amish, of course, are the best example of this phenomenon, but many other small famers practice similar economies. If not horses, they use old tractors and machinery they repair themselves, have no hired labor to pay, and often share labor and machinery with each other.

In addition to these small commercial farmers, there are also those of us who come to be called "hobby farmers," although we resent the label—believe me, what I do is no hobby, it's a way of life. But my thirty acres produce only about $3,000 in food and wood per year, hardly a commercial enterprise. If I look at my immediate neighborhood of several thousand acres, I can count about twelve part-time commercial farmers, four "hobby farmers," one full-time family dairy farm, and three very large cash-grain operations. I also count four recent failures, all larger commercial enterprises. I rest my case here: none of the small farms have been forced out by economics.

8

An Ecologically Sane Farm

1989

Driving north on Route 53 out of Kenton, Ohio, where almost 200 years ago Simon Kenton ran a gauntlet of angry Delaware Indians and survived, I ride through a level, lonely countryside settled mainly by Germans who left their country and came here in the latter half of the nineteenth century to escape military conscription. They built big, boxlike brick homes, even though they were surrounded by an absolute glut of wood. They built with brick because they were accustomed to brick, and they burned the trees for heat to bake the bricks.

The economics of government grain subsidy shapes this land now, just as once a homesick memory of Germany shaped it, and the great slate-roofed brick houses stand mostly as a monument to a past when farmers could afford to build homes that few top executives could afford today. The fences are gone now, along with the livestock and most of the woodlots of yesteryear. The great barns, once filled with animals, crumble now into a landscape of road-to-road corn and soybeans. The farmers who plant and harvest these crops spend the other half of the year in the county Agricultural Stabilization and Conservation Committee office, groveling for the subsidies that allow them to go on planting and harvesting grain they cannot make any money on without more subsidy. And so the vicious circle goes on, some surviving, even profiting a little because of inherited wealth or at least inherited solvency, and the rest slowly but surely succumbing to their indebtedness.

The month is September 1989. With a farmer's eyes, I can tell where the land is well drained and where it is not by the look of the soybean fields. A very wet spring delayed planting in these parts, and except where the soil is deep and underlain with a limestone gravel, or well-tiled and carefully cultivated, the large-scale farmers were forced to plant too late or in soil too wet, and the beans are stunted and will make barely twenty bushels to the acre if an early frost doesn't ruin them altogether.

Suddenly, out of the forestry of tall corn hiding the landscape, a narrow, paved road runs into open, well-fenced pasture fields. I imagine that I have broken the time barrier and have been launched back into 1940. A flock of sheep grazes in the field to the right, a bunch of hogs loll in a large barn lot on my left, and beyond the big brick house that dominates this acreage, cows, horses, and—what goes on here!—a herd of donkeys graze.

I brake to a stop. A lean, leathery man who could as easily be fifty as he could be sixty turns his head from where he leans on the pasture fence eyeing the livestock. He studies me as I slowly climb out of the car. This is the moment I hate, even after thirty years of reporting. Will he think I am a salesman? Will he think me too bold? Will my smile be too ingratiating? Will he be too private a person to share his work with me? Will I be able to break through the strangeness between us to an honest communication?

"I hope you'll forgive me for barging in on you, but are those really donkeys out there?" I try hard to look appealingly naïve.

"Not on this farm, they aren't," he says, not with hostility, but not with much friendliness either. "They may be donkeys someplace else, but here they are mammoth jack stock."

"There's a difference?"

"You bet your sweet life there is, when you are doing business with the mule trade."

I am helpless with ignorance. I did not know there was a mule trade in this day and age and I cannot even remember for sure what the breeding relationship is between mules, horses, and donkeys.

"Mammoth jacks are what you breed mares to, to get the big draft mules," he explains in a now kindly voice. "Mules are sterile and they're smarter than horses. They will quit working when they get tired, whereas you can work a horse to death. And a mule won't overeat and founder like a horse will. But I don't raise mules either. I just raise jacks and jennets."

I introduce myself. He is Jack Siemon, and I learn later that he is an internationally known and respected breeder of mammoth jacks. On his small, 180-acre farm he keeps about eighty head of jacks and jennets, selling about twenty a year. He can identify each of them by their looks, although he doesn't name them. "I don't think I know that many names," he says. When I show more interest, he takes me out to the field and leads me from one animal to the next, talking rapid fire, about mammoth jacks in general and his own in particular.

"Now this one here with the white mark on her cheek, isn't she a beauty? She's the daughter of that jenny we saw over in the other field. That's her sister right over yonder there, and a brother to her is in the field down the road a bit. See how she carries herself. Good heart girth, proper curve to the rump—muscle, not fat. That's as fine an example of Catalonian stock as you can find anywhere in the world. She's bred to that jack in the barn I showed you and she should have a *poppin'* good colt."

Without encouragement he launches into a short history of mammoth jacks, a remarkable animal in which is fused, he says, all the main jackass breeds of old Europe—Catalonian, Poitou, Maltese, Majorcan, Andalusian, and Italian, becoming recognized as a single breed about 1888 with the beginning of the Jack Stock Stud Book. Most of the good mammoth jack stock traces its lineage back to "General Logan," considered the best jack of the 1800s. "Some of mine trace directly back to 'Rowdy,' one of General Logan's sons," says Siemon.

At first I think that Jack Siemon has himself a nice little retirement hobby on his nice little farm, until he casually remarks that he just sold a jack for $4,000—"and I wouldn't take six thousand for that one over there." He figures that his jacks and jennets bring him, on the average, about $3,000 each, "depending." Again a casual remark catches me by surprise. He says that summer has been "one long, hard haymaking season" for him and his two hired employees, to make up for the lack of hay in the previous year's drought. It dawns on me that for this veteran farmer (he admits to seventy-six), his body still as lean and sinewy-tough as whang leather, this is no hobby and a far cry from retirement. This is serious business.

The chief "product" of his business is mammoth jacks, but they are not the only animals he raises and sells. As we walk over the 180 acres, my astonishment grows. I have been on thousands of farms from the East Coast to the West, and never before have I seen such a variety or number of animals grazing per acre: not only the eighty head of mammoth jack stock, but about a dozen draft horses, a couple of lighter harness horses, a few dairy cows and calves, a bunch of fattening steers, a flock of sheep, a barnlot full of hogs, a barnyard full of turkeys, peacocks, ducks, geese, guineas, dogs, cats, and a genetic explosion of all kinds of chickens. Every niche of the farmstead is filled with animal life, and in reaction to anything unusual, a chorus of squawks, gobbles, quacks, whinnies, bellows, bleatings, and barking erupts, all drowned out by a crescendo of ludicrous-sounding hee-haws from the jacks and jennets. Jack Siemon's farm is a celebration of the earth's vital forces.

Siemon got interested in mammoth jacks seriously right after World War II, in which he served. His wife owned a farm in Arkansas, and for a few years he tried to do the impossible: raise cotton in Arkansas and corn in Ohio at the same time. "I learned real fast that in weeding cotton, a good man and a mule could do a better and much more efficient job than a tractor weeder. But there were no good mules around. The army had bought most of them at the beginning of the war, and with the rapid adoption of tractors and trucks, mules just disappeared. So I started raising mammoth jacks to get some good mules back in circulation."

"But who buys them now?" I asked.

"Some of the Amish prefer mules, and the Amish are not decreasing. There is also a lively hobby trade in draft animals, both horses and mules. And, believe it or not, there are small farmers not of Amish persuasion who use draft animals on real working farms and in logging. In the South, mules are preferred because they can stand more heat. And remember, two-thirds of the world still relies on draft animal power in agriculture. I just had some visitors here from Colombia, for example. Nice folks. Furthermore, I still believe that petroleum will finally get so high-priced that horses and mules will again come into favor even in industrial countries. People laugh at me for that sometimes, but I'm seeing interest increase, not decrease." He pauses to let that sink in, then adds, "My interest in breeding high-quality animals actually started with poultry. I still raise and sell a few prize white Langschans. This locale was once a real hotbed of poultry breeding. At the Hardin County Fair, we'd have maybe 800 entries in the poultry show. I saw that interest come, and then I saw it fade, and now it's coming back, too, just like in draft animals." Can he account for that? He looks at me, and a grin spreads across his face. With the calculated wisdom of lean old men who have beaten the odds—even the odds against acquiring the inevitable potbelly of modern man—he says, "Oh, it's not written in stone somewhere that big-scale farming is *always* going to be the wave of the future. Could be we're in for a surprise." He pauses again, studying me for a reaction. "I'll put it this way. If you scar a tree, it heals over the wound and is stronger there than ever. That's the way a small farmer is. Owning some land is everything. You never are going to drive the small farmer out of farming if he wants to stay in. There's almost an unlimited number of ways today that he can finance his farming with a related business or job."

Siemon's "related" business is, of course, the mammoth jack stock, but more basically he brings to his farming an uncommon knowledge and skill at breeding good animals of all kinds. "I'm not a trader, I'm a breeder," he says more than once. "If you want to succeed in animal husbandry, you have to have *good* stock, whether it's mules or cows or whatever. Takes nearly as much to feed a poor cow as a good one, and to succeed, a small dairy *has* to have good cows. Now you take this Guernsey cow. The dairyman only sold her when she was a calf because she was a twin and he believed the old folklore that says a twin won't be fertile. But I could tell by the shape of her head that she was okay. I bought her and then looked up the dairyman and found she was out of a Guernsey that gave 18,000 pounds of milk a year—very good for a Guernsey. So now I'm seeing what I can breed from her. There's a fine-looking heifer calf already, and if she gives a lot of milk, she'll be worth a lot of money."

The hogs he raises are top-notch Chester Whites, the Guernsey steers are the largest of that breed I've ever seen, and the Percheron horses are sleek as silk. One draft horse Siemon raised was once grand champion at the local fair, edging out over 100 entries, and placed third at the Ohio State Fair. The sheep are a Dorset Finn cross—Finns for their reputation for multiple births, and Dorsets for being able to fatten efficiently on pasture only.

Siemon, after scratching his head deliberately, calculates that there are about 300 animals on his little farm. The common denominator underlying all of this careful husbandry is high-quality, low-labor pasture. It represents, in a manner of speaking, the third level of Jack Siemon's farming soul: after you peel away Jack Siemon the breeder and master salesman of mammoth jack stock, you find Jack Siemon the gentle husbandman, and under that one lies Jack Siemon the master agricultural ecologist. His land is a living testimony to the falsity of the modern myth that "pastures don't pay," a notion that has plunged so many farmers into financial hardship, they must depend on handouts from the government to keep them solvent while their topsoil washes away in the production of surplus grain. Except for twenty-nine acres in corn, all of Siemon's farm is clothed, in September, with bright green legume pastures that protect the soil and enrich it by taking nitrogen from the air and adding it to the soil. The oats field of earlier summer is now sheathed in alfalfa, as pretty a first-year stand after grain as I have seen anywhere. (Oats make excellent feed for mammoth jacks, horses, and sheep.)

This farm is not only being sustained, it is being constantly regenerated. After two or three years in legumes, with generous applications of manure, a field does not really need any commercial fertilizer to get good grain yields, Siemon points out. "I started farming here in 1946, and in all that time there's not an acre on the place that's received a thousand pounds of purchased fertilizer altogether, and most of that has come in the last year when I planted corn after corn because of the drought. I never really needed to use herbicides, either, but I've used a little lately, again because my rotations are out of whack because of the drought. And I'm getting old and it's easier to spray than to cultivate weeds. I'm for the happy medium. A little doesn't harm anything in my opinion."

Although Siemon's gross income does not approach that of a large-scale grain farm, I don't need a calculator to figure that his net income from this 180 acres is better than many farms I know that are ten times this size. (In fact, I know quite a few farms ten times this size that operated at a net loss throughout most of the eighties. This remarkable productivity means something more than Siemon's profits alone. It also makes it possible, and in fact necessary, for him to hire help—one more or less full-timer and one or sometimes two part-timers. Rather than invest in expensive feeding equipment, he relies on human help. It would be most interesting to punch *those* numbers into a computer. If a 180-acre farm can support two or three laborers, think how a fabric of similar farms would alleviate the problems of city congestion and joblessness and revitalize the economy of rural areas. Farms ten times the size of Siemon's, that is, of 1,800 acres, certainly do not employ twenty to thirty people. An 1,800-acre grain farm barely keeps three laborers employed half the year.

If Siemon, as a businessman, has a flaw, it is in his reluctance to find someone who can take over the farm in his absence. No one else knows the little details of how he cares for all these animals without expensive automated machinery, and even so, he finds it hard to trust anyone with the responsibility. "Yes, I should cut back. I hope someone comes along who wants to go into this business himself. There's got to be a lot of love and dedication to make this kind of farming work. You can't hire it. I'm pretty well tied to this place." Then he smiles. "But that's fine with me. I *like* being on my farm. I can't think of anyplace to go that's as nice as it is here."

9

Amish Economics

1989

The Amish have become a great embarrassment to American agribusiness. Many "English" farmers, as the Amish call the rest of us, are in poor financial straits and relatively few are making money. As a result, it is fashionable among writers, the clergy, politicians, farm machinery dealers, and bankers to depict the family farmer as a dying breed and to weep great globs of crocodile tears in anticipation of the funeral. The mourners seem to forget, or ignore, those smaller, conservatively financed family farms that are doing quite well, thank you, perhaps the best example of which is the humble, "backward," horse-powered Amish farm.

In 1985, in a speech I gave to an Ohio organization that was looking for low-cost ways to make farming profitable, I commiserated at length with the plight of financially depressed farmers. Two Amishmen approached me afterwards, offering mild criticism. "We have just finished one of our most financially successful years," one of them said. "It is only those farmers who have ignored common sense and traditional farming methods who are in trouble." He went on to explain that he belonged to a group of Amish who had, as an experiment, temporarily allowed its members to use tractors in the field. He also was making payments on land that he had recently purchased. In other words, he was staring at the same economic gun that was pointed at English farmers and he was still coming out ahead. "But," he said, "I'm going back to horses. They're more profitable."

From then on, I resolved to start cultivating the Amish as assiduously as they cultivated their fields. I had always taken our sorghum to Joe Bontrager's press in the Kenton area not far from our farm, so I knew something about the Amish knack for making money where mainstream farmers saw no opportunity. (Joe makes bent hickory rockers in winter.) We bought bulk foods and homemade angel food cake at the Petersheims' occasionally, so I knew how the Amish found ways to sell retail from their homes. We had sought advice about operating a woodworking shop at the Troyers' farm, so I was aware how the Amish were experienced in craft businesses as an adjunct to their farming. But now I expanded my horizons to include eastern Ohio, center of the largest Amish community in the world. My education in Amish economics began in earnest when I helped a neighbor haul hay to that area. If the Amish needed to buy extra feed for their livestock, they almost always chose to buy hay and raise the grain, rather than vice versa. The price of the hay was partially regained in the value of the manure, which enabled them to use less fertilizer. The manure also increased precious organic matter in their soils, a value beyond simple cost calculation. Grain farmers in my area who sold their straw and hay to the Amish were trading their soil fertility for fleeting cash gain while the Amish "banked" those products in their soil for everlasting returns.

Mainstream cultural values, or the lack of them, drove "English" society to chase after short-term profits regardless of long-range effects on society or the environment. If society as a whole drifted toward a wasteful, spendthrift, exploitive economy, it was inevitable that the agriculture tied to it must follow; abrupt dislocations of farm families in the wake of recession and inflation led to a declining rural community. The underlying cause was not a cheap food policy, protective tariffs, free trade, government subsidies, or overproduction—these were *effects*—but plain old human frailty. Most of the farmers who went under in the recent collapse of land values were, along with their bankers, advisors, and suppliers, incompetent, shortsighted, or greedy. Farmers, unaware of, or unmoved by, the fact that history has never shown us a land boom that was not followed by a land bust, got in big trouble by gambling that inflation would continue. (The same thing happened in the Texas oil boom and bust at exactly the same time.) In my county, all the farms in serious financial difficulty belonged to people who committed themselves to reckless expansionist policies. Well-intentioned social and church workers who have descended upon the agricultural scene with the battle cry of "Save the family farm" have proved ineffective because they don't recognize this feature of the farm problem—the matter of simple human frailty.

Amish culture resists financial chaos and the decline that follows by fortifying individuals against their own frailty. The culture sanctifies the rural virtues that make good farming, or good work of any kind, possible: a prudent practice of ecology, moderation in financial and material ambition, frugality, attention to detail, good work habits, interdependence (neighborliness), and common sense. Science alone doesn't guarantee the success of a farm; scientific methods succeed or fail only if they are supported by these virtues. In fact, the more familiar I became with Amish ways, the more I realized that their farming success had less to do with science and technology than it did with culture. The Amish culture provides a supportive, tight-knit community wherein these virtues can flourish.

Furthermore, Amish religion continuously challenges the merits of technological change, not only financially but spiritually. Change is viewed with suspicion, not welcomed with the Pavlovian adoration of mainstream farmers. And because they are not over-invested in it, when change forces economic dislocation, the Amish are not vulnerable. Many small mainstream businesses fail every year because they do not have this built-in safeguard.

New technology is almost as alluring as sex. Once humans have experienced any invention that eases physical effort, they will not give it up. That is the behavioral characteristic that may destroy us as a species. Curiously, only in sports do we agree to eschew technological advances, making rules, for example, to limit the power potential of baseball bats. We understand that technology will ruin our games, but we do not understand that it can also ruin cultures.

Since only hindsight can answer questions of worth, the Amish approach is to hold the new technology at bay until its true worth is known. If forbearance is ultimately relaxed in a particular case, at least acceptance will have been slowed until the good and bad effects of the new can be understood and assimilated without disrupting the community. The use of electricity offers a good example. Most Amish sects do not reject electricity anymore, but only electricity brought directly from the outer world into the home, where it may become an "umbilical cord to worldly distractions and unnecessary gadgets." Home-generated electricity, from wind, sun, or diesel motor, is generally accepted for use in the barn and workshop. In one Amish businessman's office, I found an electric lamp on the desk, powered by a wind generator. Diesel motors run electric lights, milking machines, and cooling tanks in many Amish barns, and cagey Amish boys learn to run lights from the generator up to the main barn floor so they can play basketball at night. Martin Schmucker in Indiana uses a battery-powered forklift to move big bins of vegetables from his horse-drawn wagons to his roadside stand. The batteries are charged by electricity from solar panels on the barn roof. I will bet my electric heat (but not my woodstove) that if home-produced solar energy systems for household use become practical, the Amish will be among the first to adopt them. Ironically, these systems are more practical for the Amish than for mainstream Americans because the Amish are more prepared to be content with the limited amounts of electricity these systems generate than we high-tech mainstreamers.

In farm equipment, too, the Amish are blazing a trail for the low-cost, low-energy technology that appears to be on the horizon. Their technique of pulling balers and other motor-powered field machinery with horses is often ridiculed by English farmers as fence-straddling, but the horses cost much less than the tractor power it would take to pull the machines, and of course, the dependence on fossil fuel is much less. But now some Amish innovators, like Martin Schmucker, are developing practical bullwheel ground drives to replace the gasoline or diesel engines on horse-drawn hay balers, cornpickers, and even vegetable harvesters. The planetary gearing needed to make ground drives practical required input from some of the most sophisticated mechanical design engineers in the nation. Ground-driven power shafts for many other purposes have followed, so that now an Amish farmer can use any small tractor equipment on the market by hitching it to his horse-pulled forecart. The Amish have also perfected hand-operated hydraulic lift systems for farm machinery that further negate the need for piston-driven oil gulpers.

The Amish raised recycling to a high art long before that word came into modern parlance. The amount of waste and trash produced by an Amish community is, first of all, only a fraction of that from an English community of the same size; but even so, the Amish display a genius for recycling. A conventional buggy costs about $2,500 new. Since all the parts are easily replaceable, it will last nearly forever. A century's worth of worn-out parts will not equal an auto graveyard's accumulation of just a few years. The horses, their fuel, and their exhaust, can all be recycled back into the soil—safely *and* profitably. The Schmuckers, mentioned earlier, recently built a house for their oldest married son. The dimensions were thirty by forty feet with two full stories, a full basement, and an addition measuring twenty-four feet square. Most of the wood came from buildings a nearby college was tearing down. Window and door frames were also castoffs. Making full use of their own labor and recycling, the family spent $37,500 on the new house.

The Amish are blazing trails in solving social problems, too, especially in the care of the elderly. On almost every Amish farm there are two houses. Retired grandparents live in one of them, able to carry on productive lives on the farm to the degree that they are physically able—a perfect model for today's mainstream effort to keep the elderly in their homes as long as possible. These "second homes" invariably become tiny hospitals in the grandparents' last days, where younger family members surround them with love and care. How many millions of dollars in health care do the Amish thus save, and how many millions more in their rejection of Social Security in favor of family and communal security?

There is an old Amish quiltmaker who lives near Pffeifer's Station, a crossroads store and village I frequent. Her immediate family has all passed away, and she lives now with somewhat distant relatives who, being next of kin, are pledged to care for her. Her quarters are a wee bit of a house connected to the main house by a covered walkway. I make up excuses to visit, pretending to be interested in quilts. I have no idea how old she is; her skin is smooth and crackled like old white ironstone. Around her I sense the kind of otherworldly peace I used to feel in the presence of nuns before they decided to dress up and hustle about like the rest of us. Her bedroom is just big enough for a bed and quilting frame; her kitchen is equally tiny. The boys of the family keep the walkway stacked with firewood for her stove. She has her own little garden. Children often play on her doorstep. She has her privacy but is surrounded by living love, not the dutiful professionalism of the old folks' home. And she still earns her way. Quilt buyers come, adding to a waiting list that her fingers, now slowed by arthritis, can never catch up with. But when she puts down her Bible to dicker over price, she is as canny as a lightning-rod salesman.

I love that scene. She still lives in the real world. If she were not Amish, she would have languished in some nursing home, or perhaps be dead by now, from sheer boredom if nothing else.

This economy of integration influences Amish religious practices as well. In most sects, the homes are also the churches, saving who knows how many billions of dollars. The Amish way wisely provides for the appointment of ministers by lot. No hierarchy can evolve to burden the community with yet more unproductive expense. Ministers work their farms or shops like everyone else. Their bishops do not sit in well-insulated houses in far-off cities uttering pious alarms about the end of the family farm.

Whenever I got to know an Amish farmer well enough, I asked about farm profits. Invariably, the answer was the same, spoken with careful modesty. Not as good as in the seventies, but still okay in the mid-eighties, when even the gung-ho agribusiness magazines admitted that agriculture faced a full-blown crisis. There was no critical economic problem in Amishland.

One Amish couple was kind enough to detail for me their household and farm expenses. They derive most of their income by selling milk from a dairy herd of twenty-five to thirty Guernsey and Jersey cows. On a farm of approximately 140 acres, they gross about $50,000, of which about half is out-of-pocket expenses, including taxes and interest on a farm loan that will soon be paid off. Having become relatively prosperous now, the family allows itself what would be considered frills by a beginning Amish farmer. For example, manure is scooped and loaded into the spreader with a Bobcat skid loader—their sect allows tractors in the barn but not in the fields, where they would tempt the owner to expand acreage at the risk of pushing another farmer out of business. In fact, this farm family does not work as hard as my father and I did when we were milking one hundred cows in the modern, high-tech way. We had three times the number of cows to milk (with less help to do it because, unlike the Amish, our whole family did not join in the milking), and we had to handle three times more bales of hay and straw. Ironically, we would learn, it took all the milk from seventy cows to afford the technology to handle a hundred cows, so we might as well have stuck with thirty.

The dairyman's wife showed me her expenses for July 1987, a fairly typical month, she said. There were five children in the family, ranging in age from eight to eighteen. She paid out $96.81 for food. She spent $41.89 for fabric for the clothes she regularly makes, plus some footware. Hardware supplies, such as bicycle repairs, lawnmower parts, lamp and lantern supplies, a lawn chair, cost $54.01—"It is not always that much," she commented. Shampoos, soap and other bodily care items came to $21.35. Transportation by bus and taxi (a thriving business in Amishland, by the way) came to $37.00. She spent $15.00 for anniversary cards, baby gifts, and the like. Propane, kerosene, and clear gas came to $37.50. (The family has a gas-powered stove and refrigerator, and modern indoor plumbing from a gravity-powered water system—the cistern from a barn roof sets in a hill on a higher level than the house.) They used $11.00 worth of postage stamps. Hospital aid, the Amish equivalent of hospital insurance, ran $80.00 a month. They spent $30.00 on pizza to treat the neighbors who had helped in the harvest. There was also a payment of $500 for fire insurance for the year. It came out of farm expenses, but it was not paid every year; only when the community fund into which it went needed replenishing. Some months there were long-distance calls—made from booths located along country roads in Amish communities. Phones were not allowed in the house. "You won't be tempted to use it unnecessarily if you have to walk a mile down the road to call," the wife explained. She doubted this would add five dollars a month.

The total came to about $400 a month for ordinary living

expenses. In fact, the wife thought she could "make a little" if she could convince her husband just to give her a $400-a-month allowance. "That would be a good challenge."

The family traveled far afield now and then, went out to eat on special occasions, bought books and magazines regularly. (Without television children and parents alike were voracious readers and were all able to converse knowingly on a wide range of subjects.) And there would be other long-range costs, such as house maintenance, buggy upkeep, furniture, cookware, and dishware, but "not so you'd notice," the husband said. He figured that when all expenses were totaled, year in and year out, the amount would hover between $8,000 and $10,000.

Eventually, or perhaps inevitably, I took my softball team to Holmes County for a cow-pasture doubleheader with neighborhood Amish players. It was a grand day. We were perhaps a run better than the Amish, but they were twice as adept at dodging horse droppings dotted over the playing field. Our collected "womenfolk" cheered from the shade. The Amish bishop watched from his buggy behind home plate, sorely tempted, I was told, to join the game but afraid that it might seem a bit demeaning to some of his congregation. In the relaxed atmosphere after the game, we skeptical English learned some details of how Amish farm methods result in a higher net profit percentage than any of us would have thought possible.

But comparisons of income per acre between traditional horse farming and modern tractor farming are almost meaningless because of the tremendous subsidies paid to cash-grain farmers, which the Amish, for reasons of religious conviction, will not accept. While the Amish horse farmer keeps a major share of his gross profit, the modern farm merely launders subsidy payments from the govenment right on through to moneylenders and agribusiness suppliers and rich landowners, some of whom have never even set foot on the land they inherited. The fact is that Amish farms have mostly survived, if not thrived, right on through the current economic malaise despite the fact that, by the standards of the technocrats, small farms are supposed to be on the way out.

The softball games themselves taught two lessons in economy. First, the Amish have a lot of fun playing ball without the expense of fancy playing fields and equipment. The money for our blue-and-gold uniforms could have bought a cow or two; but the Amish, in their traditional denims and blue broadcloth shirts, were always in uniform. More significantly, some of our player-farmers could not take time off from the demands of their high-tech farming machines to play in the game. The Amish had plenty of time.

10

A Horse-drawn Economy

1987

Hardly more than 200 people live in Mount Hope, Ohio, but in a way, the village is one of the greatest cities in America. There are about twenty places of business in town, and more in the surrounding countryside—with a ratio of business to population even Hong Kong might envy. Some business has arisen as a result of tourism, but surprisingly the opposite is true for the majority of it, which was there before tourism and is indeed a cause of the tourism. Nor does that count include farming, the dominant industry, or the Mount Hope Auction Barn, where between 2,000 and 2,500 head of livestock and hogs are sold every week. In addition, the weekly auction attracts scores of local vendors, selling everything from locust fenceposts to horse harnesses to garden produce out the back of pickup trucks and wagons.

Mount Hope would be remarkable enough in today's rural malaise all by itself, but its economic health is just part of a fabric of vitality that is characteristic of Holmes County and the general area around it. Amazingly, there are three other major auction houses within fifteen miles: Farmerstown, Sugar Creek, and Kidron, not to mention some smaller ones throughout the area. The four together sell upwards of 10,000 head a *week*. Their feeder pig markets are listed as far away as Wichita. Not counting forest products like lumber and locust posts, the annual agricultural production of Holmes County amounts to $62,818,000, according to statistics, coming from about 1,600 farms with an average size of 124 acres. To figure the economic impact of that kind of production, economists use a "multiplier" of 3.24 for Ohio, which means that every dollar of income generates about three more dollars as it circulates through the economy. That comes to $203,749,870 for Holmes agriculture—not bad for a county small in geographic size and with a population of only some 31,000 people.

But it is very unlikely that the economic multiplier of 3.24, while typical for Ohio, is correct for Holmes County. Holmes is the epicenter of the largest concentration of the Amish in the world, and in that fact lies the key to its remarkable economic vitality. Average cash receipts per acre in Holmes, and in adjacent Stark and Wayne counties, are topped only by super-rich Mercer, Darke, and Licking counties in west and central Ohio—with Holmes at $316, Wayne at $438, and Mercer at $531. (The state average is around $220, excluding counties with high-value truck and greenhouse crops.) But the Amish do this with low-input, small-scale horse farming and the percentage of their $316 per acre that they keep in the community is surely going to be greater than the amount Mercer County's tractors and super-duper fertilizer applications keep there. But that's not even half the story. The Amish are not only keeping more money, they are also turning it over more times in their communities. Money spent on trucks, cars, equipment, chemicals, and fertilizer in Mercer County is seldom staying in the local area, because the manufacturing is done elsewhere. In Holmes, the Amish have taken up manufacturing with the same sharp economic philosophy that made many of them moderately wealthy farming eighty acres with a team of horses. Rather than turning their farms into factories, which has proven to be biologically detrimental, they are bringing the factories back to the farm.

Any notion that a bearded Amish factory owner in his bibbed blue pants is pokey, old-timey, or uninformed about modern technology is pure myth. If you were to hear Wayne Wengerd, owner of Pioneer Equipment, Inc., in Dalton, without seeing or knowing him, you might think you were listening to a college MBA—only he makes more sense than most MBAs. When conventional manufacturers quit making horse plows, Wengerd, with the pluckiness Amish people so often exhibit, decided to start making them himself, on the strength of three years' experience working in a machine shop. "We tested the prototype over the hill, where no one could see us," he says now, with a grin. In 1977, he sold his first plow. In 1978, he quit his machinist job and made plows full time. By 1979, he was working thirteen-hour days and still not able to keep up with demand. He kept expanding until now he employs six men plus his father, who works in the office when he's not busy farming. Strangely, the increasing production is going more to non-Amish than Amish, to serious small "English" farmers who can add and subtract well enough to know that horse farming could be profitable for them, as well as to hobby farmers. "We know we have a big non-Amish market because nearly eighty percent of our forecarts go out of here on rubber tires and most Amish use only steel wheels," says Wengerd. He and his crew make forecarts, tongue trucks, farm wagons, spike-tooth and spring-tooth harrows, walking plows, and sulky plows. Despite the relative smallness of his factory, prices are fully competitive with those of large conventional factories, if not cheaper. He also makes and sells neck-yoke rings, tongue caps, wagon frames, eveners, doubletrees, and toy wagons.

To the uninitiated, a factory is taken to be the point of origin for things that are manufactured. But just as food does not come from the supermarket, so tools and equipment do not really originate in the factory. Wengerd buys raw steel from eastern Ohio steel centers and, with his hard-working crew, makes most of the parts for his machinery himself. "It means we are more in control," he says. "Doing our own machining, we can standardize quality. Also, we can make sure the parts fit so the unit goes together better than if we were buying parts from all over. Just about as important, with our own parts we know they will be available when we need them, which is not generally so true anymore, as manufacturers try to keep inventory down to a minimum. And when we have to make some of the parts anyway because they aren't available, we might as well make as many as we can. We have the equipment and the labor in place and that's half the battle." What Wengerd does not mention is the benefit this philosophy brings to the community. By starting from scratch, from raw steel, he keeps more of the labor input at home, and so the community, through his employees who do the work, becomes the beneficiary. If he were buying ready-made parts from distant urban centers, that labor-money would be leaving Holmes County. Economists do not usually figure in this kind of "profit." The sharp boys would say if the parts are cheaper to buy than make, buy them. Wayne Wengerd not only ignores that advice, but with Amish acumen, has learned to make parts on a comparatively small scale at a competitive cost, just as Amish farmers compete with megafarms. "In fact, that is why I do not want to expand much more than this," he says. "I could double my space and number of employees tomorrow, but as you add help, problems multiply. I would lose touch with the details that keep us thrifty."

The beauty of the many small decentralized factories with which Amishland bristles is how they cooperate to get the work done, forming a network of helping hands, each feeding into the other, turning the money over again and again, keeping more of the ultimate consumer dollar in the community and forming a strong interdependent web of local self-reliance. This is perhaps the most important aspect of Amish economy. Wengerd both contracts some work out to other local shops and also does some work for them. For example, he buys horse tongues from Dan Hershberger's sawmill nearby—600 of them in the spring of 1988 alone. The wood, in turn, comes mostly from local woodlots.

Tobe Hostetler does machining on Pioneer's steel wheels on his farm near Fredericksburg. "I used to provide Wayne with all his steel wheels, but there is just too much to do now," says Hostetler, who has cut and welded some 30,000 steel wheels in the last six years using diesel-generated electricity and hydraulic power. (He says his "utility bill" is the cost of six gallons of diesel fuel a week. His factory is unheated.) "I just couldn't handle all the work. We need more young people getting into this kind of business." Starting in 1966 with a repair shop, Hostetler went on to manufacture wagons and carts in addition to steel wheels, but now spends much of his time on special jobs requiring custom design and fabrication. For example, he has built a fifty-two-foot sawmill from scratch and a big log splitter that splits a chunk of wood into four parts at once. He currently employs four men.

Wayne Wengerd's Pioneer Equipment, in turn, makes parts for other factories. Wengerd made a machine to bend horse hames for a hames maker from whom he buys parts in return, including brass hame balls. He also makes parts for Alton Nisley's Machine and Farm Equipment in Millersburg, which specializes in rebuilding obsolete farm machinery, as well as for Mose Erb's E-Z Spreader Manufacturing Company on his farm near Sugarcreek.

Mose Erb uses his barn as a factory to make his manure spreaders. When John Deere, the original manufacturer, quit making its 10-A spreader, Erb, drawn to mechanics even as a boy, began rebuilding old ones, and then went on to building new ones. One of the innovations he added to the basic 10-A design was fiberglass paddles on the spreader shaft plus fiberglass guards and shields. Farmers were at first leery of the fiberglass, but the paddles have proved as strong as steel ones and are noncorrosive. "We have not had even a one-percent replacement rate," says Erb. "Everyone is very satisfied with them." He also makes forecarts and wheelbarrows. He employs four full-time workers, mostly family, and contracts out the molding of fiberglass parts to another shop. And, like so many Amish entrepreneurs, he also farms.

Alton Nisley's little factory draws even more awe. After just a brief tour of the shop, this writer is convinced that Alton Nisley and his cousin, Malva Nisley, could build the B-1 bomber, given enough time, and at a fraction of the cost modern technology claims to need. The Nisleys rebuild farm machinery, and when the old parts aren't available, they make new ones. All machinery is run either by hydraulic- or diesel-powered line shafts. No electricity.

Jonas Miller's Country Repair Shop, in Millersburg, is another small factory engaged in the same kind of work. There are more. These expert machinists even make cast-iron parts, employing the services of other shops to do the actual casting.

In decentralizing the manufacture of wooden products, the horse-drawn economy excels even more than in metal fabrication, mainly because the Amish are culturally more used to working with wood and often can rely on their own woodlots. "If America is losing its manufacturing business, it is simply because of a lack of will," says Atlee Kaufman, owner of 77 Coach Supply, near Millersburg, which manufactures shafts for horse-drawn carriages. "It would not disturb me in the least if a competitor opened a factory across the road. Americans are importing shafts for horse-drawn vehicles from Canada. I have more work than I can get done." Kaufman began his factory in 1980, after helping his father-in-law build buggies. "We could never get enough shafts," he said. When he looked into the matter, he saw one reason why. Shaft-bending machinery was unavailable and the art of bending nearly lost. Undaunted, he studied existing machines and built his own. "When we first started, there were lots of problems," he recalls, able to smile about it now. "The wood has to be very carefully selected, cured, and cut to take bending. Then it must be steamed properly and the bending done in a very exacting fashion." Today with six employees, he can turn out thirty pairs of shafts a day, buying wood wherever he can get the proper quality and curing it in his own kiln. Shafts go to individuals or to buggy shops, of which there are many in the area. But he sells nationwide—in fact, as far away as Norway.

That is just the beginning. Having learned how to bend wood, Kaufman now finds other markets for his skill and his machines, especially from the myriad woodworking shops and little furniture factories in the area. He bends bed headboards and chair backs for them—any design requiring curved wood. Such help greatly speeds up the manufacturing time for the woodworkers, many of whom are working around the clock to keep up with demand from all over Ohio and beyond.

Another rather rare kind of woodworking factory is the Holmes Wheel Shop, near Holmesville. Manager and co-owner Paul Stutzman says his factory is the largest of its kind in America and maybe the world. He employs about a dozen people and can make thirty-five to forty wheels a day. Starting in 1978 with his partner, Ron Clark, Stutzman and his crew have learned to make almost any kind of wooden wheel required—400 different sizes and designs so far. Over eighty percent of the business is outside Holmes County, so the factory is pumping lots of new money into the local area. Most of the wood comes from local sources, also contributing to the local economy. A little of the money goes out of the community, but not far. The iron hubs come from a foundry in Lancaster, Ohio, and the aluminum hub parts from a factory in Columbus.

Kaufman's shafts and Stutzman's wheels then find their way to buggy shops along with axles from small factories like Emanuel Yoder's Farmerstown Axle Company near Baltic. And if a hydraulic brake is desired, Benjamin Yoder's Valley Brake Company is not far from Baltic. Jonas Raber, owner of Raber's Cart Shop in Mount Hope, says that "Nearly everything in my carts comes from within ten miles of here." He began building carts six years ago, after retiring from farming because of health problems. He thought he'd build maybe twenty-five carts a year, but now, with a helper, he is closer to a hundred a year and has "more business than we can handle." Again, the demand is coming from non-Amish horse enthusiasts, especially draft-horse hobbyists who use Raber's beautiful models for show carts in the show ring. "Cart driving has also become a popular activity with all kinds of driving-horse enthusiasts," says Raber. "They set up driving courses through the woods and countryside—very popular."

Raber designs and builds the wood parts of the carts from scratch, using local hickory and oak, painting the carts handsomely, hiring a local pinstripe artist if the buyer wants that kind of decoration. "I think I'm the first to use a fifty-inch wheel on my carts. That makes them ride smooth. Some carts you almost need a seat belt to ride on, but you can set a cup of coffee on the seat of mine and it won't spill."

At least twenty harness shops have advertised in the Holmes County editions of Ora Miller's *Down Home Shopper's Guide,* and Miller estimates his guide lists less than half the businesses in the territory. (He got the idea for his guide when, while riding through the countryside, he counted twenty signs for home-based businesses within a short stretch of road.) Blacksmith horseshoeing shops are just as plentiful, further spreading the wealth of the Amish "automotive industry."

Many repair shops specialize in providing hard-to-find parts of obsolete horse-drawn equipment. Significantly, this business is also increasing, and most of it is from non-Amish. One of the most unique of these enterprises won't allow his name to be used. "If you write about us, it will just bring me more business, which I do not need," he says good-naturedly. "I'm overwhelmed with orders now, have had a heart attack, and just don't want any more business." So without using his name, let it be said that his business success would leave the Harvard MBAs gaping. His "factory" looks sort of like an Amish junkyard, because that's more or less what it is: a couple of acres covered with obsolete horse-drawn machinery surrounding farm buildings stuffed with obsolete parts. People contact him, usually by mail, from all over the nation, looking especially for odd and obsolete plowshares. When a visitor asks for an Oliver 110-A share, which has not been made for half a century at least, the entrepreneur who wants no more business pulls at his white, flowing beard, scratches his head, and thinks. An Oliver 110-A. Hmmmm. His mind is not what it was before his heart attack, he says. Then he snaps his fingers. "Aha!" Off he strides through a tangle of old grain binders and sulky plows into a shed that appears to be held up by various stacks of gears and wheels, turns sharply to his right through a partition, stoops over a pile of nondescript plow parts, and from a low, almost hidden shelf, triumphantly comes up with an Oliver 110-A. Asked if he had on hand as many as fifty different kinds of plowshares, he smiled and said, "Oh, *quite* a few more than that." With a helper, he also does a thriving winter business in making webs for grain binders.

The rarest business of all to flow from the horse-drawn economy is Andrew Troyer's Rope Company, which makes halters, lead ropes, hay ropes, clothesline rope, and so on. His rope is not particularly unusual, except that it is available in many different colors, but his method of making it is. Troyer operates what may well be the last rope walk in America. His "factory" is simply a line of posts across a pasture field. He and his wife or sisters walk the well-worn path beside the posts, hitching multicolored strands of polypropylene fiber to each post as they pass, and attaching the strands 765 feet away to a sled loaded with rocks. The sled keeps tension on the strands as they are wound together into a rope. Troyer adds or subtracts rocks to the sled as necessary. At the near end, the strands are attached to hooks sticking out of an old GM truck transmission and gear-shaft assembly, which is powered by a Honda washing-machine motor. This contraption can twist the strands at 1,400 RPM into 600 feet of rope at a time. Fortunately, the Troyers' rope walk is on a lonely country road. It would cause traffic jams anywhere else. "The funny part is," says Troyer with a twinkle in his eye, "I thought I had invented the thing! I didn't know that was how rope was made centuries ago."

He got the idea when his father, many years ago, found a rope machine at a public auction, capable of turning out short lengths for lead ropes. But it cost eight dollars, which his father decided was too much. So he went home and made one. Young Andrew played around with it until he had improved it into the present small rope machine he now uses— a Rube Goldberg invention he has set up in a mobile home to turn out lead ropes. "Studying this machine, I logically figured out that all I needed to make longer stretches of rope, was more space," says Troyer, "so I reinvented the rope walk. A modern rope machine that can turn out 600-foot lengths of rope without a rope walk costs forty to sixty thousand dollars, so I'm way ahead of the game. My rope walk cost maybe a thousand. At first I thought I could get by even cheaper by using the open gears off a grain binder, but it wouldn't twist good enough." With his rope walk, two people can turn out sixteen spools of rope in eight hours, while one person with a $50,000 automatic machine can turn out twenty-two. "And they can work in all kinds of weather, which I obviously can't. So I sacrifice that and the extra hand for fifty thousand dollars. Pretty good trade, I think." Besides which, that fifty thousand would have gone out of the community. The money he pays for family help stays at home. Troyer sells rope nationwide and has recently drawn interest from an international distributor.

Obviously, the Amish businesspeople are providing a blueprint for regenerating any rural area. It is interesting to apply their pattern of "automotive industry" to the piston engine society. What if we made trucks and tractors the way the Amish make buggies and wagons, keeping untold millions of dollars circulating in rural areas? Such a question does not seem outlandish after becoming acquainted with Amish factory owners. This writer is absolutely convinced that if their religion permitted it, Alton and Malva Nisley could be manufacturing trucks and tractors at least on a par with the late 1930s models—all the automotive power most small farmers need or want anyway. And this writer is also convinced that the Amish would find a way to do it at a price competitive with Detroit or Japan. Also a Wayne Wengerd could, if necessity demanded it, build the reasonably priced small grain combine that so many of us small farmers want.

There are significant lessons in rural regeneration to learn from Amish business. The decline of non-Amish rural areas is often a self-fulfilling prophecy. People are told their region is in decline until they believe it; those with ambition move away. The Amish, believing in stability, make the world bloom wherever they are. Their traditional farming background teaches them ingenuity and confidence. They believe they can do anything they set their minds and hands to. Their schools and their parents have not told them they must go to the cities to seek their fortunes, that they can't "succeed" without a college degree. They don't hear in "advanced" educational courses that small farms, small shops, and small factories aren't practical or profitable. Can you imagine an expert professor of Economics 101 telling his students there was a big opportunity in manufacturing nineteenth-century horse plows on a small scale? So the poor students spend a fortune gettng degrees that allow them to fritter away their lives in middle management, doing little productive work, but spending their energies conniving their way up the internal corporate ladders to a fat retirement fund.

The Amish entrepreneurs have only a grade-school education,

putting the lie to the biggest myth of modern society, that in education lies financial salvation. "I finally got my high school diploma," Atlee Kaufman remarked with a little smile. "As a volunteer fireman I was not allowed to operate some of the rescue equipment unless I took the test to get one." We both grinned at the irony of the situation. Here was a man with the wizardry to design and build his own wood-bending machinery, the owner and operator of a very profitable small factory, but he couldn't be a volunteer fireman without a high school diploma.

Another lesson the Amish teach, flowing directly from their philosophy of small scale, be it in farming or manufacturing, is how they integrate business into society. Most shops, even most factories, are embraced by the owners' farms. Or, at least, the owners' homes sit right next to their factories. All that separates Atlee Kaufman's house and factory is a luscious vegetable garden. Wayne Wengerd's children play in the shadow of his factory. "And when that is the case," he says, "you make sure there is no pollution." Instead of industrial parks, suburban enclaves, and huge stretches of lonely farmland, the Amish community blends all three activities into a harmonious social pattern. There is no distancing of work from family life that breeds the idea that what one does at work is not bound by the same moral code as what one does at home.

But the integration weaves its way throughout the Amish life. After grade school, learning is accompished by reading good books; job training is done not in school but in workshop apprenticeships, where it should be done. The Amish home is also the Amish church. And retirement home. Is there a computer made that can calculate the money saved in such a society, or the social concern instilled by such a culture?

11

The Barn Raising

1 9 8 3

The summer tornado that touched down in Holmes County left a path of destruction cut as cleanly into the landscape as a swath mown through the middle of a hayfield. The wind plucked up giant oaks, tulip poplars, ashes, and maples and laid them down in crisscrossed, splintered chaos through the Amish woodland. With the same nicety for borderline definition, the tornado sliced through Amish farmsteads, capriciously reducing barns to kindling while ignoring buggy sheds, chicken coops, corncribs, and houses close by. In the twenty-minute dance that the tornado performed before exiting into the wings of the sky as abruptly as it had come, it destroyed at least fifteen acres of mature forest a hundred years or more in the growing, and four barns that represented the collected architectural wisdom of several centuries of rural tradition.

But what followed in the wake of the tornado during the next three weeks was just as awesome as the wind itself. In that time—*three weeks*—the forest devastation was sawed into lumber and transformed into four big new barns. No massive effort of bulldozers, cranes, semi-trucks, or the National Guard was involved. The surrounding Amish community rolled up its sleeves, hitched up its horses and did it all. Nor were the barns the quick-fix modern structures of sheet metal hung on posts stuck in the ground. They were massive three-story affairs of post-and-beam framing, held together with hundreds of hand-hewn mortises and tenons.

A building contractor, walking through the last of the barns to be completed, could only shake his head in disbelief. Even with a beefed-up crew, it would have taken him most of the summer to build this barn alone and it would have cost the farmer $100,000, if in fact he could have found such huge girder beams at any price.

The Amish farmer who was the recipient of this new barn smiled. The structure, complete with donated hay, grain, and animals to replace all that was destroyed by the storm, cost him "about thirty thousand dollars, out-of-pocket money"—most of that funded by his Amish Church's own internal insurance arrangement. "We give each other our labor," he said. "That's our way. In the giving, nothing is lost, though, and much is gained. We *enjoy* barn raisings. So many come to work that no one has to work very hard. And we get in a good visit."

The outsider listened, dumbfounded. The barn raising had already shaken his faith in the religion of Modern Progress in which he had been raised. He had come to see a folksy rural skill of the nineteenth century and, instead, witnessed a practical example of how to survive rather elegantly in the modern world.

The first day, the Amish installed the girder posts, girders, sills, joists, and flooring over the lower level where the livestock would be housed. The oak girders were sixteen inches square and fourteen feet long, hoisted up on the girder posts by human muscle heaving in perfect unison. The joists, half the size of the girders they rested on, were mortised into the sill beams over the foundation. Floorboards were laid down over the joists. Then the Amish carpenters installed the horse stalls, cow stanchions, calf pens, bullpens, pigpens, mangers, feed boxes, hay and straw chutes, all with so lavish a use of wood as to make a cost-conscious modern builder weep.

While this work was in progress, the most skilled carpenters were sawing to size the timbers that would become the post-and-beam skeleton of the barn, then marking and cutting the mortises and tenons by which the posts and beams would be put together into "bents." (A bent is the basic structural unit of a post-and-beam barn. It consists of at least two vertical posts connected by two horizontal beams, with additional braces notched in at each corner for greater strength and rigidity.) The mortise holes were first bored round with brace and bit, then squared to size with mortise chisels and corner chisels, the work moving along rapidly. All the while there was steady conversation about the Yoders' new baby, the price of horses at the Kidron auction, the possibility that the Stolfuss family might not make it from Indiana to the raising in the morning.

"The raising draws all the attention," one of the carpenters told a watching outsider. "But this is where the real work is done, measuring and cutting the joints accurately. The raising is just putting pieces of a puzzle together."

The procedure for joining timbers properly has changed little for three centuries. A hole is bored through both mortise and tenon and a wooden pin, of white oak or black locust, is driven through the hole to lock the two beams together. The pins, before use, are dried in a little makeshift kiln kept fired nearby. Once it is driven into a joint, the dry pin swells slightly in normally humid air, while the greenwood beams dry and shrink slightly over time. The resulting bond is so tight that, even after a century, the pins will sometimes be impossible to drive out. For joining wooden structural members, no method has impoved upon this classic mortised, tenoned, pinned joint.

There was no detailed plan of the barn construction, although the building was large and complex. "The blueprints are right up here," a carpenter said, pointing to his head. "Not so difficult as you would think. There's a standard way these barns go together and the overall design does not change much from farm to farm. The size of the barn determines the number and dimensions of the bents, and the dimensions of the bents rule the dimensions of the posts and beams. It was all figured out long ago."

By six-thirty the next morning, a traffic jam of buggies clogged the country road to the barn raising. The main bents were all laid out in proper order on the barn floor, ready for raising. Every Amish male who could swing one carried a hammer, and they stood around in expectant little knots talking quietly as the sun poked up over the cornfields. In the house rose the sound of female voices, the warmth of their chatter alone enough to start the great pots of food to cooking. "You mean to tell me there will be a barn standing here before the sun goes down?" a visitor asked in disbelief. "Oh, yes," one of the carpenters replied. "Fact is, you'll be able to put hay in it by noon." The visitor laughed, thinking he was being teased.

At precisely 7:00 A.M., the head carpenter, or "boss of the

raising," as he is called, shouted in German the traditional order to begin, and without fanfare, seemingly with great casualness, some twenty bearded farmers poled the first bent into upright position, while twenty more held it with ropes from falling on over the edge of the foundation. Just as unceremoniously, twenty more workers quickly poled the second bent up and immediately the more agile of the young men climbed up the beams like monkeys, jiggling and fitting the connecting beams into their proper mortises, and driving in the locking pins. The barn raising was under way.

The casualness was deceptive. "I couldn't sleep a wink last night," the boss of the raising admitted later. An older man nodded understandingly. "*Ja*. You know I had to quit bossing on doctor's orders. Too hard on my blood pressure."

What appeared at first as a rather distracted and unplanned bustling about on the part of several hundred workers was, in fact, an operation being run with almost military precision. Under the boss of the raising were two assistant bosses, and under them was a group of men recognized for their skills in particular departments of barn construction. Each of them headed up a crew, while the majority of the workers simply joined a crew according to the type of work they felt most comfortable doing, or where they saw another hand was needed. Skill and sometimes age determined the choice. Nimble, younger men worked high in the framing, fitting the beams together. Older men mortised and built doors. Strong men hoisted up rafters and beams. Little boys gathered up waste wood and piled it out of the way. Besides the many crews of polers, siding nailers, roofers, and rafterers, there was a special group framing doorways and windows, another soldering and hanging spouting, another putting together a new hay track out of the wreckage of the old one, to be hung later high under the roof peak across the haylofts.

All this work went on simultaneously at various sections of the barn. Yet few orders were given. The men knew what to do. The boss of the raising and his assistants and crew captains merely orchestrated the flow of work, like band directors leading skilled musicians. The barn grew, organically, in one cacophonous symphony of whining saw and pounding hammer.

One worker was using a skill saw powered by a gasoline motor. Sometimes the motor would not start, and he would glare at it. When it did start, he grimaced at the noise and smoke it produced, clearly uncomfortable with this rather un-Amish tool. Why, he was asked, did the Amish go to such great lengths to avoid electricity when the gas-powered replacement seemed religiously just as repugnant? The farmer pushed his hat back and toed the saw, now lying on the ground. "The Bible says we must not be yoked to the world," he explained. "The electric, it *yokes you*. This gas motor—I can take it or leave it. And to tell the truth, right now I'd just as soon leave it."

Another bystander who heard the question wanted to help answer it. His mother was originally Amish, he said, and was shunned by Amish society when she married an outsider. "They even had a funeral service for her. Her parents never spoke to her again." But hc was not bitter about that, he said. He respected the Amish. "Electricity *is* a yoke, you know. They think if they let it in their houses, the temptation would be overwhelming to get all the gadgets electricity encourages, like the rest of us do."

The issue reminded him of an experience he had when he was younger. "I used to work with my Amish kinfolk. I helped in the threshing. The bishop had a big lug-wheeled steam engine to run the thresher, and every time he drove it to another farm, those steel lugs would tear up the road and the county commissioners would raise hell and send him a bill. One day when I arrived at the threshing, there sat the steam engine with the lugs removed, replaced by rubber treading over the steel wheels. I always kidded my kinfolk about their ways, so I said, 'Heavens! Rubber on your wheels, Bishop! God's gonna getcha now.' He laughed, but took it kind of serious, too. 'Well, we prayed over the matter and studied it a long time,' he explained. 'We finally decided that it was not the rubber *itself* God was against, but *riding on air.* Only angels should ride on air!' "

Around the edges of the little army of Amish workers gathered the outsiders—the "English"—most of them brandishing cameras. A television crew set up to record a scene to lighten the evening news.

"Unreal," the photographers kept murmuring as they clicked their camera shutters. The Amish elders stayed busy asking the Engish not to take pictures. The English stayed busy trying to sneak a few anyway.

"But what is wrong with taking a few photographs?" some of the bolder photographers protested.

"A picture leads to pride," the elders tried to explain. "It is against our religion."

The Amish and the English engaged in a staring standoff then, exuding mutual bewilderment. The English could not understand a religion that viewed images of reality with suspicion. The Amish could not understand a religion for which the image was the reality.

The head carpenter's noon prediction was wrong. The barn was ready for hay by eleven, an hour ahead of schedule. He nodded with delight when the English visitor who had laughed at his prediction apologized. "We'll be finished by three," he said, and this time the outsider did not laugh.

The speed of the raising was not attributable just to the large number of workers. A good third of them were standing around talking or eating at any given time. The secret was that the men not only knew what they were doing without being told, but they always knew what to do next. The work flowed. The workers were also extremely "handy," a word the head carpenter liked to use. Because of their lifestyle, the Amish knew how to use their hands, their whole bodies, in physical work. They could perform physical tasks in less than half the time and energy it might take a typical office worker, or even a typical blue-collar worker trained to do only one job well. Watching the Amish workers, one observer said he no longer believed it necessarily took one hundred thousand slaves twenty years to build the Great Pyramid, even if it was 482 feet tall and 775 feet square. With a bit of practice, a hundred thousand Amishmen could have built it in less than five years, he decided.

In addition to individual handiness, the Amish farmers were taught by their tradition how to work efficiently together. At the tedious task of nailing on the hundreds of siding boards, for example, there was no eager elbowing for room on the wall, each man intent upon seeing how many boards he individually could nail down. Instead, crews of older men, away from the barn, marked each board with a chalk line indicating the place where the board crossed the beam it would be nailed to. Then these men started nails along the chalk line. Other men quickly carried away these boards, invariably three at a time, and passed them up to younger men clinging to the beam frames. They, in turn—one above, one below—slapped the boards in place with one hand and drove the preset nails in with a hammer in the other hand, the siding going on in almost a continuous wave, as if it were being slowly unrolled.

By three o'clock, the barn was finished, even to the hanging of the doors. The workers hitched up their buggies and went home to tend to their livestock. By 7:00 P.M. they had all returned, this time with wagons laden with hay and grain, pigs in crates, horses and cows in tow on ropes, to fill the barn with the feed and animals the tornado had taken away. When that work was completed, more food was served amid convivial rejoicing—as close to a party-like celebration as the Amish ever come.

Two English farmers, leaving the party to which they had been invited, walked silently to their car. The summer night rolled quietly over Holmes County on the wings of fireflies. One of the farmers finally spoke. "Makes you wonder if some of them folks might not consider praying *for* a tornado once and a while."

12

Not So Friendly Persuasion

1985

Last summer, Henry Hershberger built a new house on his farm near Apple Creek, but instead of winning society's approval for having achieved one of America's most cherished goals, he got two weeks in jail. Hershberger nevertheless continues to have faith in the house that has caused him so much anguish. He built it himself with the help of skilled neighborhood carpenters, so he knows it is built to last. He invigorated the old foundation upon which the house rests with ton upon ton of new concrete. He installed special flashings designed to keep termites at bay. He used full two-by-four studs of oak. The exterior siding is lapped redwood, and the roof, channel-drain metal. "We were going to use shingles on the roof," explains Hershberger, whose ruddy, bearded face and constant smile make him look like Santa Claus, "but shingles only last twenty years, you know."

Though the house is brand new, one walks into it a hundred years ago. Kerosene lamps shed a warm glow over the interior, softening the hard, straight lines of its spare furniture and austere decor. The only obvious compromises with physical comfort are the two bentwood hickory rockers with thin cushions on their seats. Dominating the kitchen is a massive woodburning cookstove, so stout that Hershberger, himself a bit stout in his middle years, can hunker down on the open oven door and its hinges do not give a fraction of a millimeter. Amanda, his wife, sits at the kitchen table, which stretches at least twelve feet across the spacious room. She is packing brown eggs in cartons. The floors are oak hardwood throughout, stained a beautiful dark brown with linseed oil. And spotless.

A cloister-like quiet pervades the house. There is no electricity, no televison, no radio; the only music is the dulcet chiming of a pendulum clock striking the hours, a clock Amanda believes is well over a hundred years old. There are no rugs on the floor. The unadorned walls are solid white plaster above and gray wainscoting below. Along one wall, on a low bench, sit four children, round-eyed and wraithlike in the lamp glow, as naïve and innocent as a primitive painting. But when they think no one is watching, they whisper in each other's ears and giggle soundlessly.

The only plumbing in the house is a faucet from which spring-water issues under pressure from a hydraulic ram that has operated continuously for twenty-four years. Hershberger wants his visitors to taste his water. Among the Amish, who avoid more spirited liquids, waters are judged much the way gourmets judge wines. The visitors sip. Nod their appreciation. There is no foxy taste of sulfur in the water, just enough lime to make it dry but not too dry, and a delicate infusion of iron to give it full-bodied flavor. They all agree. Henry Hershberger knows his waters.

Thus fortified, Hershberger sits down with his visitors by the stove in the living room, surrounded by his family (there are four boys and four girls, but the two oldest girls are away teaching school now) and begins to tell the story of how he went to jail. Hershberger is a minister in the Swartzentruber branch of the Amish Church, the strictest sect of all. It is, in fact, more correct to say that his religion, rather than his house, landed him in jail, as even the building-code inspectors admit that, with the customary exemptions allowed the Amish, the house passes muster with no problem. But Hershberger refuses to even *apply* for a permit, such application being, in his conscience, against his religion. The bureaucracy is prepared to look the other way when Hershberger ignores those parts of the code that do not apply to Amish houses, but for Hershberger not to apply at all, not to offer token obeisance to authority—that is unforgivable. Ronald Rehm, acting prosecuting attorney in the case, says as much. "We've bent over backwards trying to compromise with these people. We've waived all the points of the regulations except our right to come in and inspect for violations."

"But if you have waived the parts of the building codes the Amish won't comply with, what is there left to inspect?" he was asked.

"Well, the structural factors," replies Rehm.

"But haven't you already admitted that the Amish build better structurally than anyone else?"

"Oh, yes. They are good builders. The best. But the law is the law."

Obdurate to all compromise, Hershberger was found guilty of violating the permit regulation, but given thirty days to obtain a permit. The authorities did not understand the commitment of their gentle-seeming adversary. When he did not obtain a permit in the prescribed time, Judge Ralph Amiet fined him $5,000. When he refused to pay the fine, the judge sent him to jail to work it off at twenty dollars a day. Two weeks later, Hershberger was abruptly freed with $4,720 of his fine still unpaid.

The sheriff was ordered to confiscate enough of Hershberger's property to satisfy the fine. "They came and counted all the animals and wrote down the numbers on their paper," recounts Amanda serenely. "My hens, though, they are so wild. They fly this way and that and are very hard to count. So they gave up and left that number blank on the paper." She laughs, proud of her wild chickens, which do not have to subscribe to nonviolence as she does.

But the sale of property never took place. Local auctioneers refused to cry the sale. "I said I wouldn't receive stolen property, not even from the sheriff," growls Chester Horst of Apple Creek, a Mennonite who has become Hershberger's most vocal supporter. "They said they could legally force me to do it, but I was of the opinion they could only force me to make my facilities available. I said they could use my auction barn anytime I wasn't using it, but there was no way anyone could force anyone to cry the sale. They never proceeded. Public opinion was running too strongly against them."

For a man most people in Wayne County believe is being persecuted for his religion, Henry Hershberger seems diffident, insulated from such adversity by an impenetrable serenity, though he says his health has suffered from the tension. He displays neither rancor over having gone to jail nor elation at having been freed.

He is not eager to justify his actions or explain his position. The conflict between the permit business and his God's business is clear beyond words to him, and beyond his power to control. He is not comfortable speaking in English, anyway, even in his home, surrounded by friends, let alone in court.

"Oh, I try once to explain it," he says, haltingly, always searching for the right word. "I try to tell the judge at my, my, what you call it . . . not trial . . . not hearing . . . *ja!*—at my 'arrangement.' But I not so good talking, and the judge, he gets angry at me and he says it's not *either* a matter of religion. So I keep still. If he thinks he knows more about my religion than I do, what can I say?" He shrugs and sighs. "I go to jail."

Another Amishman in the room, David Kline, speaks up. "The Amish have been going to jail for four hundred years for what we believe. We are against violence and so the only way we can fight for our freedom is to resist passively and go to jail if we have to. Nonviolence is more effective, anyway." A hint of a smile plays at the corners of his mouth. "Some of the boys were saying the other day that it might not be so bad to spend some time in jail. They would get to watch television for once."

According to Hershberger, the controversy began one day last summer when he was nearly finished with his new house. Working in the basement, he was suddenly aware of a man standing behind him—the building code inspector. "He not announce himself before coming, not wait at door to be let in," recalls Hershberger. "I was upset over that. He told me I . . . I . . . break law for not getting permit. I asked him if he knew what a door was for. He said he didn't want violence. I said there would be none. I would carry him out as gently as I would a baby."

Nothing much except talk has transpired since Hershberger

was released. Another judge has fined two other Amishmen who refused to apply for permits, Eli Hershberger and Dan Yoder, $600 each. Their attorney appealed the case to the Ninth District Court, leaving the whole matter very much up in the air. But public outcry—mostly for, but also some against, the Amishmen—continues to seethe and bubble. The Wooster *Daily Record* received so many letters to the editor that it finally refused to print any more until there were new developments. The letters are a study in communal examination of conscience, probing deep into the ambiguities of American society. The first letters state the obvious two sides to the argument: religious persecution versus "the law is the law." The next wave of letters begins to grope for a workable definition of religious freedom that would determine whether religious freedom was being used only as an excuse to avoid governmental red tape. But as the letters bark back and forth at each other across the editorial page, what ultimately emerges is not an argument about Hershberger at all, but a debate about whether building codes, adopted in the sixties, are, in fact, necessary, effective, or even constitutional.

"Hey, now you're getting to the real point," says Chester Horst at his sale barn. "What we're facing is a question of everybody's freedom. At first it was just a matter of paying twenty-five dollars for a permit and go build your house. Even the Amish went along with it. But this permit business keeps getting stickier. A permit isn't automatic. Now they [the County Department of Health] want to test your soil. People already in a community get notions they don't want any more people coming in. The Amish think things have gone too far. There's a distinct possibility their children will not be able to build new houses on the ancestral farms. There's too much regulation," he says, "and the Amish understand that."

In the quiet glow of lamplight, beside his stove, his children listening, Hershberger is asked what it is like to be an Amish minister in jail. "A long time between meals," he replies. "But there were five men in my cell. Talking made time go a little faster. The language in there you don't want to repeat.

"They had to take my picture. I said it was against my religion. I kept my head down. The nurse, she stuck up for me. But the sergeant, or so they called him, he came in and said he was going to have my picture and if I didn't look in the camera he'd see to it I did. Well, I looked down, but he shoved his arm against my forehead and forced my head up." Hershberger smiles with the patience of 400 years. "I guess the Lord won't hold that against me, will He?"

13
A Patriarch Passes

1990

It is not unusual to see Junior Frey's team and hay wagon clip-clopping through Kirby, since that's the shortest way to get from his farm to his father's place. But the wagon Junior Frey's son drives today carries a coffin, not hay. It is his grandfather's last ride. Slowly the wagon leaves the little Catholic church of St. Mary's at the edge of the village and heads down the road toward the cemetery just visible across the plowed fields. There is no sound save the soft blubbers of the horses, the knock of hooves on the road, and the flinty grinding of ironclad wooden wheels. Ninety-one years earlier, the man in the coffin, Jerome Frey, had been carried out of this same church on his baptismal day. In all that time, the landscape has scarcely changed—the same uncluttered country road, the same plowed fields, the same farmsteads on the horizon, the same huddled quiet of village houses. Bearded young men serve as pallbearers. Horses stamp their feet. Another old farmer rides that final mile in a horse-drawn wagon. The only hint that change has come to Kirby is that the farm fences are gone, and the cows that the fences kept out of the village are gone, too.

Then something unexpected, unplanned, happens. Before the cars with the widow and her eleven sons and daughters can move out to follow the wagon, a crowd of younger people stream from the church, striding out into the road to keep pace with the wagon. Grandchildren and great-grandchildren mostly, but a mixture of friends as well. Some carry babies or hold children's hands. A true patriarch is passing. He leaves seventy grandchildren and 116 great-grandchildren. And he leaves Kirby.

Jerome Frey would have been pleased to know he would go to the grave leading his family as usual by gentle persuasion. "By hickory," he would have said, using his favorite expletive, "this is just like it oughta be." Likely as not, he would have repeated some old folktale, like the story of Abe Rosenberry's funeral at Bowsherville, a town not far away, once bustling with shops and people, though long vanished now. "They all got drunk at Abe's wake," Jerome Frey would have recalled, "and never noticed when the coffin slid off the wagon on the way to the cemetery. Dug the grave and kept on drinkng. Prayed and preached for the poor man only to find he had somehow gone off and took his coffin with him. Took most of the night to find it in the dark and then pray him on to heaven all over again."

Kirby might well have gone the way of Bowsherville were it not for Jerome Frey. Kirby is a village of maybe 200 people when the Corner Pocket and the Blue Room are open, but it's still ringed by a good-sized farm population, not the vast, empty reaches of large, absentee-operated grain blocks.

Kirby is the kind of village that the hand-wringers, all over America, say is dying today. Kirby thrives largely because Jerome Frey and his family made it thrive. He always said he was "just a farmer," but he understood that good farming was not very profitable in itself. "You gotta put back in the soil what you take out so it stays fertile," he would say. Rather than plunder the soil and use the profits to buy out other farmers, chasing them away to the cities, Jerome always had at least one other nonfarm business going. And he taught his sons and daughters that such diversification was the key to economic survival: "Don't put all your eggs in one basket."

When he was younger, he cleaned clover seed, ran a thresher, and operated a sawmill. He shipped fruit, hogs, and beef by rail out of Kirby. In winter, he hauled ice from farm ponds to the seven saloons Kirby sported in those days. "Ice was about the most profitable crop we raised," he said with a smile. "Nature bore the full cost of making it." Later, to start his sons in business, he bought the hardware store in Kirby, then the lumberyard, then started a construction and roofing business, and finally bought the grocery store. When his son Bud got sick, Jerome stepped in, though he was far past retirement, and worked the grocery store himself. It was not because he needed the money, but because he hated to see Kirby without a grocery.

If there had been a gifted student program when he was in school, Jerome would have been picked, and it would have ruined him, filled his head with nonsense and packed him off to college to become a white-collar wonder conniving for tax-financed tenure and retirement at fifty-five. But there were only eight grades in his day, although for two more years he went back after the crops were in to take what his daughter Aggie Thomas refers to as "postgraduate eighth-grade" work. He had a genius for teaching himself to do anything he wanted, from the extremely tricky job of reshaping a bent sawmill blade to playing the violin. He passed his handiness on to his offspring. They all grew up believing that they could do anything. Then they would do it. Operate their own small businesses. Build their own houses, barns, and furniture. Organize their own musical groups. Paint their own pictures. Write their own poems.

Jerome developed a formula for self-reliance in his community that his offspring and a few of his neighbors followed: own a little farm for subsistence and stability, do your own building and repair, operate a small business, and work long hours. He'd get one after another of his own children started on this formula and then he'd step back, suppressing his own ego so they could learn by experience. "You don't want to be too bossy," he said. "A young man at work has to be free to make his own mistakes, too." It was pleasure enough for him to watch the family and Kirby prosper.

Even when the rural economy faltered in the eighties as it had in the thirties, Kirby and the Freys flourished. Jerome had taught them how to live to survive bad times. "We learned during the Depression that you can live real good without much money," he would say. "Make do with what you've got. Provide your own food and shelter. And your own fun. My wife played the piano, and I played the violin, and the young people would gather above where the Blue Room is now and dance. No one could have had more fun than we did."

The Freys were one of the first to have a car in the community and one of the last to quit farming with horses. "Progress is okay, but we throw the old out too fast. You can't tell young people you could still farm a hundred acres with horses and make a living today, even though the Amish get rich that way. I always think how I used to harvest all the grain around here with a two-thousand-dollar thresher that is now harvested by half a dozen hundred-thousand-dollar combines. The only difference is, they get done a lot quicker now. You can't tell a young person how much nicer it was, compared to driving, to get on a comfortable train right here in Kirby and take a trip in style. Trains were so wonderful."

But he seldom traveled. Went to Florida once, he said, and nearly froze to death. When asked a few months before his death if there was anywhere he'd like to go, he said no. Just wanted to be around home. Wasn't there anything he'd like to see before he died? He thought a little bit and said, "Well, I wouldn't mind watching a plow turn over rich dirt once more."

And now we who loved him, young and old, family and friend, follow the wagon to the cemetery, remembering. We glance at each other and see a resolve blazing out of our eyes into the sunny raw air of this early spring day. The torch has been passed. By hickory, we are doing what he taught us, and we will teach our children. Get a little land free and clear, even if you have to work three jobs to do it. Learn to make do for yourself. Build self-reliance, without which this country is going to go completely to hell. By hickory.

I had to travel thousands of miles, live in eight different places in six different states, spend half a lifetime in schools and libraries listening at the knees of scientists and artists to learn finally what Jerome Frey just naturally knew: I should have stayed home. He understood that his world here around Kirby was the whole world; that he could learn the secrets of the universe here as well as anywhere else; that the only escape from provincialism was to know your own province deeply enough.

Following Jerome Frey to his grave, I see that the high-heeled shoes of a young woman walking ahead of me have worn a hole in her silk stocking, and her heel is bleeding. She had not come to the funeral prepared for a walk to the cemetery. She pays the blood no mind. I do not think she even knows. I look down at my own feet. It has been a long time since I have walked over horse slobber splattered on a road. And even longer since I'd watched tears falling behind wagon wheels.

14

A Woodcutter's Pleasures

If one has cut, split, hauled and piled his own good oak, and let his mind work the while, he will remember much about where the heat comes from, and with a wealth of detail denied those who spend the weekend in town.

—Aldo Leopold

1988

It took me ten years to learn how to cut wood, but nearly thirty before I understood why I cut wood. I was walking up through our grove one December evening, glancing up and down the trees I passed, my mind absorbed in identifying each of them, assessing its condition, deciding its fate. White oak, a beauty, clear trunk to twenty feet, maybe thirty inches in diameter, possible veneer tree . . . but would I ever bring myself to cut it down? White ash, fungus on trunk, dying, at least two cords of firewood in it . . . but hole in top could be horned owl's nest, so won't cut that one either. Hickory, dead, must cut soon . . . hickory rots quickly when exposed to the weather.

As I ruminated in this fashion on my way home, suddenly I beheld what looked like a bank of flickering votive candles in a church sanctuary. I stopped dead in my tracks. Was I hallucinating? I often thought of my woodlot as a cathedral, but this was a bit much, even for Thoreau. The lighting display quickly explained itself. As I had walked along, one of my stacks of firewood had become silhouetted between me and the dying sun, and each space between the split sticks of firewood in the rick was glowing with its own individual little window of sunset. I thought at first of running for a camera, but realized the apparition would be gone long before I got back. I stood and enjoyed. I wondered if the sight would ever repeat itself in my lifetime. So many conditions had to coincide to allow it to happen: the sun at that particular position, the precise time of year and day, a woodpile stacked as sloppily as this one, and at the correct angle to catch the sun so uniquely, clear weather, and my footsteps at just the right distance from the woodpile. I watched the vigil lights fade and knew why I cut wood.

There have been other such moments: the rare evening in November when the red rays of sunset turned the dark tree trunks to glowing orange; the time I thought moon rays had pierced the night floor of the forest, only to find, on close inspection, phosphorescent toadstools gleaming on a fallen branch; or the drab March day when a quick rain squall mixed with hail coated the firewood I was splitting and stacking, leaving a glaze like wet varnish on new furniture. The squall abated, the sun broke through the scudding clouds, and suddenly the ends of the red oak and white oak sticks I had stacked in the rick absolutely glowed reddish orange and golden yellow. Off to the side, a tangle of wild black raspberry canes glistened with a patchwork of shiny purple in the sunlight. What made this sudden bloom of color so delightfully noticeable was the hail that had fallen and, for a brief time before it melted, covered the ground with sun sparkles.

As I sat on a stump and thought about these sudden, unpredictable displays of beauty—the chance to meditate is not the least of the pleasures of woodcutting—I realized that it didn't take me thirty years to notice such marvels. What had taken so long to learn was that these seeming trivialities were important events in my life. Until I understood that and realized that in journeying across a woodland grove I could experience wonders as exciting or exotic as I might find in journeying across an ocean, I did not understand why I persisted in cutting wood when I could afford more slothful ways of heating. Woodcutting is an easy and basically free entrance into a natural world that becomes more exotic as one's knowledge of it deepens.

For example, last fall I discovered a deer scrape—a small patch of ground pawed bare by a whitetail buck. I was awestruck. It was an open love letter to any doe who passed by, saying: "I shall return." The deer left his scent on the scrape and, as bucks often do, a print of his hind hoof—his signature—in the middle of the bare patch. And up above the scrape, I found the usual branches, with one twig broken and dangling. The buck had broken the branch with his antlers or teeth and then rubbed scent on the branch stub from a gland in the corner of his eye to arrange a tryst. The doe would come along and rub her eye gland scent on the branch, too. She thus reads her lover's message and leaves her own: "I am waiting."

By trial and error mostly, I have learned that woodcutting is pleasurable for me if I follow certain rules, which are far more important than the how-to of woodburning experts speaking the language of BTUs. (Always remember when someone is attempting to impress you with British Thermal Units that you can lift 700,000 of them with one arm—that's about the output of fifty pounds of hickory wood.) How these rules apply to work in general is easy enough to see.

* * *

1. I try never to rush anymore. Chain saws and falling trees kill woodcutters who get in a hurry. Besides, nothing done in a hurry is ever done well. If work is not worth doing deliberately, it's not worth doing at all.

2. I don't push myself beyond my limit of physical endurance. Woodcutting kills or injures tired people more often than it does hasty people. Haste and exhaustion together are partners in crime: they make a fine formula for a bloody gash if you are young, or a heart attack if you are older. "Slow and steady go far into the day," the old farmers said—and add pleasure to the work, I might add.

I delay tiredness by varying work activities. Instead of continuous sawing, continuous splitting, or continuous stacking, I do a bit of each in turn: saw off a section of tree trunk, split it, rank it in the stack, proceed to saw off another section. I take time out to inspect with binoculars (which I always have handy) the redheaded woodpecker who is cussing me from the top of a nearby tree. There's not a flashier bird in all of the tropics.

3. I cut and split no more than five cords a season. As a rule of pleasure, I believe that fun decreases as cordage increases beyond that amount. If you need more wood than four to six cords a year, then (a) your house is too big; or (b) it is not insulated well enough; or (c) you are not operating your stove properly; or (d) you have poor-quality wood; or (e) you are burning wood early or late in the season, when above-fortyish temperatures usually mean some other type of backup heating would be more efficient and save you having to cut more wood.

4. I never use a power tool when a hand tool is better. Most of the woodcutting injuries I hear about come from trimming small branches with a chain saw, especially when the woodcutter is in a hurry or tired. Small branches—up to two and a half inches in diameter—are easier and faster to trim with a sharp ax. Also quieter. You can even hear tufted titmice calling their incessant "Peter-Peter-Peter" when you are swinging an ax. On twigs less than an inch in diameter, I use a heavy lopping pruner. For splitting wood, I use an eight-pound splitting maul, never a power splitter. Heavy-duty power splitters are too expensive to justify for five cords annually, and the cheaper ones make the work harder than hand splitting, not easier. Use the heaviest maul you can comfortably lift on hard-splitting wood. On kindling or very straight grained wood that splits easily (like red oak), a lighter ax does fine and is less tiring. In splitting a section of tree trunk, the natural tendency is to split the way you cut pieces of pie—in wedges. Often wood won't split that way easily. Instead, split around the outer circumference of the chunk. An added advantage of this method is that the chunk will stand upright longer as you split around it, rather than compel you to bend over and set up the chunk again after each split as when splitting pie wedges. Place the chunk to be split between two others, or against a log, and it will stay upright while you split off most of the pieces without any other help.

5. I don't use a hand tool when no tool is better. Crotch pieces that are difficult or impossible to split by hand I burn in our big fireplace, use for boiling off maple sap outdoors, or save for woodworkers. (Crotch pieces often have beautiful feather grain.) Another old farmer's saying: "If you don't use your head, you have to use your hands and feet."

6. When I start getting tired from harvesting wood, I grab pruning saw and loppers and cruise through the woodlot, doing what I call "playing God," my favorite woodland pastime. Where a clearing in the woods (from a downed tree usually) has allowed

sufficient sunlight, or where old fields are reverting back to woodland, tree seedlings will sprout and grow thickly. Here in mid-Ohio, oak, ash, maple, elm, hickory, black walnut, wild cherry, and occasionally hackberry, sassafras, mulberry, and others will volunteer this way. As a general rule, the tree that gets its head above the others in a given area will eventually take the sunlight away from those around it and they will die. So I cut the seedlings down around the ones I want especially to grow up.

7. In the interest of pleasant woodcutting, I will not wallow around in deep snow. It is a good way to lose wedges, numb wet hands, or dull chain saws cutting into ice and froze ground. (I don't know why ice dulls chain saw blades worse than wood, but it does.) It is even less pleasurable to cut wood in hot weather. The time to enjoy woodcutting is from October to Christmas and from about Groundhog Day until May 1 here in mid-Ohio, barring heavy snows. Sometimes January is nice enough, sometimes not.

But don't let cold weather alone keep you out of the woods. On windy winter days, the lee side of the woods is often surprisingly pleasant. When cold wind blows from westerly directions, I'll work a tree on the east side of the woods, and vice versa.

8. I rarely cut down big trees. Felling them is dangerous work, and all those directions in books (including mine) don't begin to tell you everything you need to know. If you have a big tree in sound condition that needs to come down because it is dying, sell it. Timber buyers are taking about anything now, even cottonwoods. The professionals will cut the tree, haul away the primary log they buy, and leave you the rest to cut for firewood. Or perhaps you need some lumber yourself. For a little more money, a sawmill that saws lumber for customers will cut down the trees and haul them, too.

Another way to avoid felling large trees is to wait until they die and let the wind blow them down. I do this most of the time, which causes my local forester to want to pull his hair out except that he is even balder than I am. To a forester, the proper way to manage woodland is always the way that makes you (not to mention the timber buyer) top dollar. Foresters don't know what to make of fools like me who don't care about top dollar. I don't mind letting a tree grow old, rot at the base, and fall down in the wind. My reason is that old dying trees have holes and hollows in them for squirrels, raccoons, owls, bees, a host of woodpeckers, nuthatches, wood ducks, opossums, and a billion or so insects and microorganisms without which the woodland could not function properly. I persisted in letting stand a big old black oak snag even though it meant watching about a cord of wood slowly rot away, because such dead trees are much loved by flickers and woodpeckers for nest holes. To my delight this year, I spied a pileated woodpecker drilling on the snag, the first time I have ever seen this uncommon bird in our county.

Quite often when a hardwood tree finally blows down, I can saw off the bottom two feet of rotten trunk and still have a good sawlog, plus plenty of firewood from the branches.

9. I think of woodcutting as a game, and my "work" as a chance to improve the skills involved. Lopping two-inch branches accurately with one blow of the ax may not be quite as satisfying as placing a line drive over the first baseman's head from the right side of the plate, but the satisfaction is similar in kind. In splitting, I used to have to resort to steel wedges all the time, but now only rarely. Splitting a chunk of wood is like cutting a diamond—both materials have their fracture planes—and the more you learn about various species of wood, the cleverer you become. If all else fails, let the stubborn chunk freeze and then split it. Some kinds of elm

will hardly split any other way. Nature is trying to tell us something, I think. White elm is poor firewood. Willow is even worse. It not only burns cold, but stinks as well.

10. Stacking wood is artwork, too. I thought I had learned to make a good rick until I saw my neighbor's woodpiles, ranked on an otherwise empty barn floor. All the numerous stacks were as straight as if laid up with a level, each kind of wood in its own rick, also separated according to size—from mere twigs to pieces roughly three inches in diameter, and not one speck of dirt on any piece. Because his stove takes only small pieces, the sticks were barely a foot long, which lent a uniquely neat appearance to the ricks. I gasped at the man's attention to detail. This was not manual labor; this was sculpture.

Certainly for me, the satisfaction of looking at my ricks of wood ranks right up there, pun intended, with strolling through an art museum, even when the sun is not turning them into banks of candlelight. They are my monuments to fuel independence. The first year my stacks all fell over when the ground thawed, but that is now a rare occurrence. Every person has his or her own idea about how to build a rick. The only rule is, do not try to stack another woodcutter's wood, any more than you'd try to adjust that person's stove. Some want little, big, and medium-sized sticks piled together; some do not. Some want self-standing ends; some want to build their ricks between trees. Some woodcutters' ricks tend to lean away from them as they stack, and some tend to lean toward them. Each stacker must learn his or her weakness and make adjustments accordingly.

Almost every stick of wood will be fatter at one end than at the other. By studying each piece as you place it in the rick, you can keep the top level. Actually, when you finish off the top layer, it should not be level, but tilted slightly so rain runs out to the edge of the rick, not down through it. Many woodcutters cover their outdoor stacks with tin or plastic roof panels. Although I often do so, because I have a pile of scrap roofing, I have learned it is not necessary. Top off the rick with two rows that are slightly tilted, the top row covering the cracks between the sticks in the row under it. For topping off, I like to use wide, thin pieces that come from splitting around the circumference of large log sections. If they are laid on top of the rick as described, and stick out from the rick a wee bit on the down side of the tilt, the water will run off nearly as well as if the rick were covered with tin. Sometimes when I have large slabs of bark handy, I will use them to make a topping for the rick.

11. I keep a half-dozen sharpened chain-saw blades on hand. The chain saw is the most unpleasant part of woodcutting because of its noise, and a chain saw with a dull blade is pure exasperation. With a dull blade, you have to listen to the noise twice as long. If you change blades as soon as they stop throwing off big fluffy flakes, you will save your saw for many more years of service and cut twice as much wood in a day. "Better six blades and one saw than one blade and six saws." No old farmers have said that yet, but give them time.

12. If burning wood is to be a pleasant way to keep warm, you must have dry, cured wood. Nothing turns a person away from heating with wood faster than trying to start a fire with wet or green wood. Also, when you replenish a fire, dry wood will blaze much more quickly with much less smoke. So I cut two years ahead of use. What I cut in the winter of 1988–89, I will burn in the winter of 1990–91. I transfer wood from the woods to the garage before snow flies or between snowstorms so there is no snow on it and any rained-on pieces have dried before the wood is burnt.

I may save money by burning wood, but I doubt it. I listen to people brag about how many BTUs their stoves put out in the same way I listen to people brag about how many miles per gallon their cars go. I start to yawn. I do not burn wood to save money, but to stay warm. No other heat keeps me warm unless I turn the thermostat up to ninety. If I figure costs based on using electric heat with the thermostat set at ninety, then, of course, wood saves me money. Burning wood, I am free of thermostats. The peace of mind I feel, knowing that when the electricity goes off at ten below zero my stove and my wood provide the warmth, the cooked food, and all the hot water I really need, is worth every splitting blister I get.

The very act of watching wood burn brings another pleasure that probably only some deep probing into our evolutionary development could explain. I am thinking now not only of the contentment to be had watching a log burn in the fireplace or flicker through the window of a stove, which is considerable, but also of burning brush in late winter. If someone would pay me minimum wage to stack and burn brushpiles, I would do it forty hours a week and be quite content. You should not really burn much brush—leave the piles for birds and wild animals to hide in. A certain amount of twigs should be cut into bundles for starting fires, too, or for quick heat if you are cooking on your stove. But occasionally I find myself with brushpiles where I don't want them. The trick is to burn them where you won't set the whole woods on fire. It's best when there is a little snow on the ground, in a cleared area away from trees or burnable debris. If you have stacked the branches closely and neatly, they will burn even if green or wet, as long as they're started well. Then all you do is lean on a fork, gaze into the fire, dream, and be warmed by its flame, occasionally forking the outer, unburned branch tips in upon the hot coals and keeping an eye out for a burning spark to waft beyond your cleared area. Don't ever burn on windy days or if the forest floor is dry. Brush burning time is perfect for a picnic—the fire extends a little circle of summer around it, but without bugs. I use a stump for a table and log sections for seats. I stare heavenward at the trail of a jet and give thanks I am where I am.

Economy (and therefore satisfaction) is gained in woodcutting in direct proportion to how deeply one immerses oneself in woodland culture. Roughly three-fourths of the United States was naturally a forest. That suggests a lot of things we do not usually think about. We are, or should be, a woodland culture, with an economy based heavily on wood. Many of our ecological problems are the result of denying this truth. We have been imposing a grain and oil economy on a natural woodland that could produce more food and energy, at least on sloping land, with trees. The land rebels with erosion, with climatic conditions that swing from one extreme to another, with water shortages and surpluses, with humans unhappily hungering for the tree grove environment they evolved in. If enough land could be taken out of cultivation and put into stored solar energy (woodland), wood would be our primary renewable source of energy and our principal raw material for manufacturing. We even have the technology to harvest wood nearly as easily as we harvest grain or oil. We could be using more of it, not less, in housing, furniture, tools, automobiles, sports equipment, and recyclable paper products. Wood only costs more now because it is becoming scarce. In the natural forest areas of our country (the eastern two-thirds especially) the most important environmental contribution a person can make is to nurture a twenty-acre woodlot. Anyone who really wants to can buy a woodlot with money saved by avoiding expensive vacations and weekends, overpowered cars, or oversized houses. If your woodlot eventually becomes the location of your home, you may very well pay for it with what you save by not buying a building lot.

Deliberately placing myself within a woodland culture, I find myself automatically following a saner home economy. The wood is there, a free by-product of my investment in the land. I am not really harvesting firewood; rather, I am improving a woodlot for eventual sale of valuable lumber. The firewood is a by-product. The ashes, returned to the land, enrich it. More than that, my son is a cabinetmaker, and what we are *really* doing (I help when I can) is creating the raw materials of his business. But I think to myself, as we work together, rolling a walnut log onto the truck, that what I am really, really, *really* doing is enjoying the companionship of my family. Who can put an economic value on that?

I think about all these pleasures as I lean against a tree to catch my breath in December twilight. It has been a singular day. I spotted a bluebird, a species supposed to be miles south of here this time of year. But the wild grapes, dried to raisins now, the poison ivy berries, the bittersweet, the hawthorn fruit, and the multiflora rose hips, which at other times I am inclined to cuss, are still there to feed on, so the bluebird remains. From where I stand I can see the kitchen lights through the trees and my wife moving about, getting supper ready. A soft snow has begun to sift down through the darkening air. I smell food cooking and woodsmoke. Soon I will be basking my tired bones by the stove. I think of Robert Frost's poem "Stopping by Woods on a Snowy Evening." But I do not have "miles to go before I sleep." And maybe that is the woodcutter's greatest pleasure of all.

15

The Pond at the Center of the Universe

1991

The man standing stone-post-still on the shoreline of The Pond was watching a muskrat swimming on the water surface, its wake forming a V-shaped ripple of scarlet fading to indigo against the sunset. Without turning his head, which might scare the muskrat into diving underwater and scooting for its den, the man also watched, out of the corner of his eye, a great blue heron drifting down out of the sky toward him. He was used to seeing the heron on its nightly trip up the creek valley, headed back to the rookery where most of Wyandot County's herons, silent and solitary by day, gathered to roost. But this time, the huge slate-gray bird, its wingspan over five feet, was doing something wary great blue herons do not normally do. It continued to drift down in the twilight, made a pass over the pond, and then turned straight at him as if to land on one of the posts that held the homemade pier he was standing on. Forgetting the muskrat, but still not moving a muscle, the man watched aghast as the great bird hovered above him, like an avenging angel, and perched *right on top of his head.*

Not many people would have the steely nerves to suffer, without moving, a great blue heron's talons gripping his head, but this man, my brother-in-law, is not known in these parts for reacting to anything in an ordinary manner. He had already realized that no one was going to believe him unless he caught the bird. He started inching his right hand up the side of his body. Slowly, slowly, slowly. Gotcha! With one swift grab, he snatched the heron's legs in his hand like a chicken thief removing a hen from the roost and bore his prize homeward so that all the neighborhood might see and believe. His family gathered round, ignorant of the danger involved. None of them knew that great blue herons can skewer an unsuspecting human's eyeball right out of its socket with one lightning stab of its beak. This time, fortunately, its captor wore glasses and when the heron jabbed at him, it only knocked the glasses from his head. When another onlooker reached for the glasses, the heron speared him in the hand, having endured, it seemed, enough human attention for one day. A quick decision was reached. In the case of herons, better two in the bush than one in the hand. The bird haughtily stalked away, looking like the dignified old lady who hoped no one was watching when the wind momentarily blew her dress over her head. Then it regally pumped its wings up and down, slowly lifted itself into the air and flew away.

Life at The Pond, as we have always called it, has been full of such adventures. One of my sisters, who lives where she can see The Pond out her window, watched a cormorant repeatedly waddle out on the diving board, raise its wings slightly in the typical way cormorants do, and dive in. My sister assumed the bird was diving for fish, but she was not sure. "It looked exactly as if it were just having a good time," she says. "And who knows. Maybe it was."

Another day, I arrived at The Pond for a hockey game to see about a dozen little nieces and nephews sprawled out, face down on the ice. My first thought was that they had finally done it—killed each other in one grand hockey massacre. Closer examination, however, revealed that they were peering down through the crystal-clear ice at a trio of snapping turtles, their carapaces as big as meat platters, clearly visible scarcely two feet below, lolling on the pond bottom as if it were June. If we all laid there without movement, fish would congregate under our bodies, obeying an instinct to hide under logs, which they now mistook us for. The ice had become a giant television screen, tuned to nature's own PBS station.

Three generations of our family have worked, played, fought (the only verb that properly describes our hockey games), picnicked, swum, camped out, made out, and celebrated holidays around The Pond. Most of all it has been a haven where any of us could come when the need to be alone hit us, to sit and slip out of the consciousness of self and into the arms of a little wilderness that thrums and hums with enough activity to keep a naturalist occupied for a lifetime or two. It is not an accident that Thoreau gained inspiration for his best nature writing on the shores of a pond.

A pond, surrounded by meadow and with a grove of trees growing nearby, attracts and concentrates an amazing diversity of wildlife. In this humdrum corn-belt country of north-central Ohio, The Pond has hosted, by my count, over forty different kinds of wild animals, not counting hockey players. In addition, we have identified at least 130 bird species around, on, or above The Pond. I have not begun to learn the names and numbers of different insects, the most fascinating pond wildlife of all. There is a little water bug, for example, by the name of *Hydrocampa propiralis,* which likes to eat the leaves of water weeds. However, it can't swim or is too lazy to try, and so, like a good American, it uses technology instead. It builds itself a tiny boat out of bits of leaves, and sails off into the wild blue yonder.

As I play Thoreau and watch the life of this little wilderness in action, there evolves in my inner vision, a scene of seething, roiling, dynamic consumption. The Pond is an endless, entwined, labyrinthian dining table, at which sit the eaters being eaten. Barbaric as that vision seems, it is the accurate view of nature, a view without which ecology remains only a vague word, incomprehensible to both environmentalists who wish to protect nature and entrepreneurs who wish to subdue and exploit nature. The Pond teaches that life is not so much a progression from birth to death but a circle of eating and being eaten, the chemicals of one body passing on to form another. A frog becomes a charming prince or vice versa, not by a kiss but by the magic of the biological chain of life.

I sit on the bank and peer into a clump of cattails. As Yogi Berra said: "You can observe an awful lot, just by watching." A muskrat chomps on the rhizomes at the bases of the cattails. The rhizomes are good for humans, too, if cooked like potatoes. (The young pollen spikes, steamed, are offered as gourmet food in fancy restaurants, and "ears" of this "cattail corn" sell in specialty West Coast supermarkets.) However, the muskrat must enjoy its delicacy with one eye over its shoulder, watching for mink, for whom muskrats are a delicacy. (Muskrats make good human food, too.) The mink, in turn, had better be alert for the great horned owl nesting in the woods next to The Pond. The owl is not at all deterred by the odorous oil the mink can unleash, skunklike, when disturbed, and Mrs. Great Horned Owl's young would appreciate a change in diet from the red-winged blackbirds she has lately been snatching off the cattails where the birds roost.

Sunfish hide among the cattails, where they hope the big bass will not find them. The sunfish look for snails to eat, which in turn are feeding on algae. If the sunfish watch out only for bass, they may not notice the little Eastern green heron standing like a statue at the shoreline, ready to grab and gobble them. And if they elude both bass and heron, the kingfisher, which sits on the dead branch of the shoreline hickory tree, may dive-bomb into the water and spear them. If the kingfisher is not around, beware of cormorants practicing for the Olympics.

The algae, meanwhile, compete with the cattails for nutrients in the wastes dropped by muskrat and heron and redwing. A frog sits among the cattails, too, half hidden by them and half by its own camouflage colors. The frog doesn't know that the cattails and the other pondweeds protect it; they just make a convenient place to hide while it waits to snatch flying bugs attracted by the pondweeds' flowers. On the upper stalk of a cattail, a dragonfly perches, waiting patiently to make dinner of a mosquito buzzing by. In the pond, dragonfly larvae feed on mosquito larvae, while fish feed on both. Attached to the cattail stalks under water, often in symbiotic nutritional relationship to them, are diatoms and blue-green algae being eaten not only by snails, but various insects and worms. Other types of algae—filamentous algae drifting in the water—become food for bullfrog and toad tadpoles. The snapping turtles, themselves being parasitized by leeches sucking their blood, will eat some of the bullfrog tadpoles, and I will eat some of them after they grow up to be frogs. And, by and by, I shall eat a turtle, too.

Even those plants and animals that die a "natural" death—the most unnatural death of all—do not escape the feast at nature's table. Bacteria eat decaying matter on the pond bottom, and produce ammonia. Other bacteria "eat" the ammonia and turn it into nitrites. Still other bacteria turn the nitrites into nitrates. The algae and plankton then eat the nitrates and turn them into proteins. Microscopic animals eat the microscopic plants, and the proteins, carbohydrates, and minerals in them begin the long climb up through the biological food chain. If that sounds complicated, understand that I am oversimplifying the process exceedingly.

The largest or most cunning eaters at the head of the table are kept from destroying the whole food chain because they are the most vulnerable to changes or shortages in the menu—as the dinosaurs once proved. The exception to the circle of diners is rational man, who is clever enough to find sustenance in almost any part of the food chain, but who also has the chilling freedom to rise above it, to act against nature. Thus, man, when he does not understand the full impact of his awesome powers, becomes nature's greatest danger.

As I watch, The Pond becomes a giant magnet attracting the wildlife around it. The barn swallow skims the surface of the water for bugs, the raccoon and opossum fish from the shoreline, the deer come down to the water for a drink, a black rat snake basks in the sun, having already raided a redwing's nest and satiated itself. A cedar waxwing flutters above the water for bugs. A wood duck floats on it, diving for food. A flycatcher darts out over it and back again to a tree. A buzzard soars high above it all, watching me, hoping that my stillness means I'm dying and that it can get to me before the undertaker does.

My father built The Pond in 1950, with his little Allis Chalmers WD tractor and its hydraulic manure scoop substituting for a bulldozer. Whether he knew it or not—not having any technical engineering experience in such matters—he picked an almost perfect spot for a farm pond. About a fourth acre in size, The Pond drains water from hardly a ten-acre watershed, almost all of it in woodland so that no silt-laden water from cultivated farmland, saturated with fertilizers and toxic chemicals, can wash into it. Most amateurs want to build a dam that catches the runoff from many more acres than that, which means a larger pond, which is hard to take care of properly, and an expensive mechanical spillway, plus an even larger emergency spillway to keep water from overflowing and washing away the earthen dam. The Pond has no pipe and concrete box spillway at all, only an emergency grass spillway, off to the side of the dam, which, despite expert opinion to the contrary, has proved to be all that is necessary, barring some catastrophic flood. Our kind of pond, using an earthen dam to hold the water, is called an embankment pond. The other kind of artificial pond is referred to technically as an excavated pond, easier to build and maintain because it is hardly more than a big hole dug in the ground. Many excavated ponds can be found in northwest Ohio on generally level ground, or dug into big hillsides in the more mountainous southern regions, or along highways where the soil was excavated for roadbed construction. Embankment ponds are more characteristic of gently rolling country where small ravines cut through low, but relatively steep, hills.

Neighbors stopped by during the summer construction work on The Pond to speculate about when, or if, it would fill with water. Uncle Ade finally reached a verdict.

"It won't fill up for two years," he wailed. He always talked as if he had to drown out the roar of a grain harvester.

"And it'll dry up every August," Uncle Lawrence chimed in.

Dad, who did not get along very well with either Lawrence or Ade—or, come to think of it, anyone else—paid no attention to them as he scooped the dirt out of the ravine and pushed it into a pile for the dam. The Pond was overflowing by Thanksgiving.

The Soil Conservation Service (SCS) agent also stopped by to cast dubious eyes on the project. He offered to survey and design the dam properly, which Dad took as a kind of effrontery.

"Don't need government help," he said.

"But if you let us design it, the government will pay a third or more of the cost," the technician explained.

"Yeah, and if I do it your way, it will cost me three times as much, too," Dad snorted.

As a result of Dad's stubborness, our dam leaked a little, not having the prescribed clay core in the center of it. It had a number of other minor flaws, too, but with some repairs now and then where muskrats tried to dig holes through the dam, it has neither washed away nor gone dry in forty years.

Dad would not take the wildlife experts' advice on stocking the pond either. The standard stocking practice in those days was so many largemouth bass to so many bluegills, which, time would prove, led to a pond overpopulated with bluegills that did not grow to practical size for either eating or catching. (Nowadays, you can stock your pond with hybrid bluegills that are sterile or with channel catfish that do not usually propagate new generations in ponds.) Dad put only largemouth bass in The Pond, big lunkers he caught someplace else. Anyone who hooked one of those lunkers had to throw it back in. "You'll see," he said."They will cannibalize most of their own young, and so we'll have no overpopulation and the remaining bass will grow to good size." The wildlife experts sniffed their disapproval and left. One day in 1957, several of us caught ninety-eight bass, all of a pound or more in weight, which seemed to us to settle the argument in Dad's favor. (About that time we did heed the advice of the SCS and planted "living fences" of multiflora rose on the farm, which eventually turned into a living hell of briars. I have not much listened to expert advice since then.)

The SCS had more luck with citizens less contrary than my father, which would include nearly everyone except me. Even if its multiflora rose and autumn olive are turning many Ohio pastures and woodlots into one huge living hell of briars (someone has suggested that there might be some good come in it—the multiflora in eastern Ohio may eventually stop the flow of East Coast garbage into the state), the Service can boast justifiably that it has designed and helped build an estimated 9,000 artificial ponds and small lakes in Ohio. No one seems to know how many artificial ponds there are in the state altogether, but SCS officials in Columbus estimate about 10,000, counting those not assisted by SCS, "and maybe more." The federal pond-building program was probably the most beneficial government effort for the public good ever funded, so, naturally, it has been all but dropped. Farm ponds can, in some cases, slow the runoff of water to rivers, alleviating the effects of floods and decreasing the amount of soil erosion. The water that is held back may recharge groundwater or evaporate back into the atmosphere to recharge the hydrologic cycle. The ponds further benefit society by taking out of cultivation land that often should not be farmed anyway, ravines and hills that profit-squeezed farmers would otherwise be planting in erosive row crops. Also, ponds near houses and barns can be used for fire protection, and thus lower insurance rates. Many ponds, particularly in the northwestern part of the state, as around Defiance, are used as a source of house and drinking water to get away from high-sulfur well water. In many parts of the state, particularly in the southern hilly regions, ponds were built for livestock water even before the SCS came into being. Some ponds are also used for irrigation purposes.

But although Dad was full of ideas for growing fish commerically in our pond, and using the water, enriched with fish manure, to irrigate a super-duper market garden below the dam, The Pond has been used only for recreational and social purposes. It has been the symbolic, if not real, center of our family's activity. All nine of us siblings still live in this rural county, six of us and our families more or less clustered around The Pond. Thousands of other Ohio ponds have served the same purpose for other families—a close-at-hand vacation spot and health spa. The only accessory The Pond lacks that can be found at other ponds is a cabin beside it. Next to The Pond lies what might be mistaken for a large lawn were it not for the bare spots that mark the bases and the pitcher's mound. In summer we play softball, in fall football, and then the scene shifts to the ice for the hockey wars. In addition to the ubiquitous pier and diving board, The Pond has a small area of concrete apron on the shoreline where the third generation waded and played as tots, free of the mud, and with a sandpile beach above it. Steel posts, set in the concrete apron to hold the net that kept the children from wandering into deep water, stick dangerously above the ice during hockey season, and we have talked for years about cutting them off. But now a fourth generation is coming along, and the posts may be needed again.

The muddy bottom is the bane of all farm ponds, and the best way to avoid it is with a pier and a raft anchored in the middle of the pond, which is what most people do. My cousin, who happens to own a stone quarry, decided he would dump a few loads of crushed stone into his pond to make a mudless bottom. Ton upon ton upon ton he dumped. The stone just seemed to disappear into the bowels of the pond. The beach he made needed almost yearly additions of sand, too. What sand did not sink mysteriously into the earth ran off into the water and then sank mysteriously into the earth. He finally realized that, even with his own stone quarry, it was costlier to try to turn a farm pond into a swimming pool than to build a swimming pool.

His pond, known as Eagle Park for at least three-quarters of a century, was the scene of the original hockey wars before they moved two miles away to The Pond. The wars at Eagle Park involved the whole community and were dominated by Dad's generation, especially Uncle Lawrence, who flew up and down the ice on racer skates, the blades of which were eighteen inches long. To keep playing after dark, we soaked straw bales in oil and burned them, one behind each goal. One afternoon, Uncle Lawrence, bored with hockey, drove his old, wooden-spoked truck out on the ice and skidded around in giddy circles, whooping like a kid. He finally overdid it and slammed the truck into the bank sideways so hard that one of the wooden wheels snapped in two. Undaunted, he fetched a tree branch out of the grove and wedged it in under the axle to serve as a sort of sled runner in place of the wheel. Still whooping, he drove the jalopy home that way.

Eagle Park in the 1920s and thirties really was a park, with a baseball diamond as well as the large pond for fishing and swimming, and a nice grove of hickories for picnicking. Across the road from it was a little red-brick schoolhouse, crumbling away today, which I suppose was the reason the park came to be there in the first place. Grandfather Rall built the pond, and he did not do it for frivolous reasons like providing the community with a park. He needed water for his sheep. He drained the pond once, and even though I was only a little boy then, I can still remember big catfish wriggling down the sheep paths below the dam in a couple of inches of water, looking as out of place as a bishop on a manure spreader. Grandfather grazed sheep not only on the pasture around the pond but on the ball diamond, too. Waste not, want not, he said. Eagle Park had its own baseball team, which had the reputation of whipping all challengers. Fritz Cassel, who attended that little red schoolhouse and would later serve twenty years in the Ohio legislature, says he was the water boy. Literally. "Whenever a ball was hit into the pond, I had to go after it," he says.

Since nearly everyone in the neighborhood was a farmer then, whenever the ice was thick enough, work stopped, whether it was midweek or weekend, to play hockey. *Weekend* was not a word in our vocabulary. I wonder now, after all these years of progress and prosperity, how many rural neighborhoods have a free park all to themselves, kept manicured at a profit by sheep, and with the time to enjoy it? When we were poorer, we were a whole lot richer.

The hockey wars shifted to The Pond largely because Dad installed lights that allowed us to follow our madness far into the night without fear of running into burning straw bales. And it continued for a while to serve as a community watering hole as well as a family gathering place, just as Eagle Park did and many ponds in Ohio still do. The Pond hosted lodge meetings and church groups (although not immersion baptisms as Eagle Park once did), and especially school parties. These parties moved from pond to pond, depending mostly on which owner had young people at the right age for such activities. But gradually, most of the ponds have become, for the time being, anyway, forgotten little domains of wild nature where only those with old memories go now and, long to, as James Whitcomb Riley said of his day, "Strip to the soul and dive once more into the old swimmin' hole."

But if the countryside empties of people, it fills with more wildlife. The deer that sift out of the woods for a drink at The

Pond were unknown in the days when catfish swam down sheep paths. Canada geese, once very rare in these parts, have become a destructive pest around farm ponds. And even ten years ago, no one would have thought a cormorant possible here, let alone working on his half-gainer. Life is a wheel forever turning. Whatever goes around, comes around. I have a hunch even the young people will come back someday after they realize the cities have deluded them.

But one wild species, *Homo hockiatis,* is definitely dwindling, being found in all the county only on The Pond, and then only in reduced numbers. Will a whole crowd of them ever flock again to The Pond to beat each other with hockey sticks or sit on the bank by the fire drinking hot chocolate? I think the peak year was 1957, when even in February we were all still eager for one more game. Snow had fallen six inches deep on the ice, however, and a warm wind was melting it and the ice. Uncle Lawrence decided the only way to remove the wet snow quickly was with our Allis Chalmers and its manure scoop. Dad did not think much of the idea, but he went along with it. After all, this probably would be the last game of the season.

Since the tractor had little traction on the ice, Lawrence would start out in the grass, careening along in road gear till he got to The Pond, then drop the blade and let the weight of the tractor slide it and a scoopful of snow to the other bank. Halfway through the job, halfway across the pond, the thawing February ice gave way, and the tractor sank four feet into the water, while Lawrence sat astride the seat, whooping hysterically. "You're still the craziest man I know," Dad yelled at him, shaking his head. "*Me* crazy?" Lawrence roared with hyena-like laughter. "This is your tractor, not mine!"

16

My Wilderness

In human culture is the preservation of wildness.

—Wendell Berry

1990

I used to say that it was but a few steps from the world of my garden to the world of wild woodland, but now I realize how that statement reflects one of the most invidious errors we humans have been making.

It certainly is true that my garden borders woodland. It is also true that a pronounced change in my mentality occurs when I slip from my workaday garden into the wilder haunts of the woods. I am transformed from Mr. MacGregor worrying about Peter Rabbit into Tarzan rallying the jungle animals against the excesses of human civilization. Nor would I deny that my garden serves the side of my rational mind that demands MacGregor-like order in a chaotic world, while my woodland provides me with the wilderness that the mystic, wild side of my nature yearns for.

The error is in thinking that these contrasts represent different worlds. Vegetable gardens are perhaps more human-controlled than are wild woodlands, but the difference blurs with close scrutiny. Every effort to impose an order that would sever the garden completely from wild nature ends in silly futility or catastrophe. One year a neighbor of mine decided that, by God, he was going to get rid of every weed in his sweet corn patch once and for all. He drenched the soil with atrazine above the recommended rate. No weeds for sure, but nothing else would grow there either for three years. At the other extreme, we preserve "wilderness areas" as if we could store nature away like a can of pickles to satisfy momentary cravings. I went to a wilderness area once and got trapped in a colossal traffic jam. The only wildlife I saw was elbow-to-elbow campers emitting mating calls from portable stereos.

If gardening has taught me anything, it's that we can't separate ourselves from wild nature. Even in a hydroponic greenhouse I recently visited, a cat was kept to control mice, and shipments of ladybird beetles were unleashed to eat the aphids. We live in union with a wilderness fundamentally beyond our control or we don't live long at all. We don't have the choice of moving from a human world to a nature world, but only from one footstep to another. As Theodore Roszak put it so well in *Where the Wasteland Ends* (1972):

> We forget that nature is, quite simply, the universal continuum, ourselves inextricably included; it is that which mothered us into existence, which will outsurvive us, and from which we have learned (if we still remember the lesson) our destiny. It is the mirror of our identity. Any cultural goods we produce

> which sunder themselves from this traditional, lively connection with the nonhuman, any thinking we do which isolates itself from, or pits itself against, the natural environment is—strictly speaking—a delusion, and a very sick one. Not only because it will lack ecological intelligence, but because, more critically still, it will lack psychological completeness. It will be ignorant of the greatest truth mankind learned from its ancient intimacy with nature: the reality of spiritual being.

I had to step back and forth from garden to woodland many times before I realized that the line between them was too fine to draw, that the "reality of spiritual being" dissolved the difference I had imagined. Amid the jungle-like fernery of the asparagus patch, for example, nature plays out dramas of eating and being eaten as wild as those that occur among the bulrushes of the woodland creek: the chipping sparrow flits from her nest in the strawberry patch to prey upon larvae of asparagus beetles with all the grisly intensity of the black rat snake snatching into its gaping mouth a field sparrow bathing at the shoreline of the creek. Wren battles wren for territorial rights to thc birdhouse in the apple tree as ferociously as two bucks in the woods battling for supremacy of a deer herd.

The difference between the larvae of ladybird beetles attacking aphids on the lima beans and cheetahs attacking wildebeests on the Serengeti Plain is one of scale only. I learn to measure my progress as a gardener not by the size of my tomato harvest, but by the degree of calmness I can maintain when I abruptly meet a garter snake hunting slugs.

There is only one accurate way to describe the roiling, moiling, toiling scene of the healthy garden: *it's wild!* Hundreds, perhaps thousands, of species of bugs, birds, worms, and animals move in and out of it, all eating and being eaten. Yet most of the time, this banquet table of soil provides enough food for me, too. The real need to "protect" it comes only when nature's normalcy has been thwarted, either by its own seemingly chaotic workings or by that of humans.

An ecology-minded world would not need to protect gardens from rabbits because gardeners would understand the continuum of nature and ensure the natural habits and habitats of owls, hawks, foxes, and other animals that feed on rabbits. All else failing, humans would eat their rabbits themselves, with the same gusto that they eat Big Macs. Cabbage patch and wilderness would be one. Tarzan understood gardening better than Mr. MacGregor.

I walk from one part of my property to another as through a continuous wilderness. The vegetable rows, the woods, the pasture, the creek bottom, the little grain- and hayfields are all "garden." They are all part of the Great Garden that once covered the Earth and might cover it again. As I walk, I pass only from one realm of the Great Garden to another. The more indeterminately the borders coalesce, the more assuredly I achieve the oneness of the natural continuum. The vegetable garden, the most humanly shaped realm, becomes a kind of decontamination chamber, a place where I can slough off the fretting cares of civilization while I pull weeds—lamb's-quarter, purslane, pigweed (wild amaranth), and sour grass—some of which I realize, wryly, are nearly as tasty as the salad plants I grow.

Then I step into the woods by way of a glade that also serves as backyard lawn. I leave the yard deliberately unkempt so that the mower freaks who visit me can't tell where lawn ends and wood begins. Who can say whether I should mow here or not—whether I am obeying the strictures of lawn neatness that our rural middlewestern mentality teaches? Raspberries at the woodland edge further blur the border between civilization and wilderness. Are they part of the garden or the woods? I ask the same question of the hickory nuts hovering over them.

In the woods I become a sort of high-tech Tarzan. Loincloths unfortunately are not approved of by rural middlewestern Germanic souls of propriety any more than unmowed lawns, but my belt holds a knife and more (magnifying glass and hand pruners). With binoculars around my neck, I can watch for what food, spiritual or corporeal, this wilder garden has to offer today. I find a luna moth—an endangered species in this region, where even woodland is sometimes mowed—newly emerged from its cocoon, still not ready to fly, glistening pale green and purple. I hold it in one hand and study it through the magnifying glass with the other. I am transfixed by its beauty. Of the unlimited arrangements of color and pattern that moth wings could take, why these particular ones?

I am face-to-face with mystery I cannot fathom, appearing over and over wherever I turn my eye. I begin to understand the meaning of "reality of spiritual being." Here is knowledge that science has not yet imagined, not visible to magnifying glass or the most powerful microscope. The moth flutters away. It soon will mate and lay eggs if a bird does not catch it first, and then it will die shortly, its magnificence "wasted" if not for my chance meeting with it. *Perhaps* wasted. In the realm of spiritual being, *perhaps* is the most necessary word in any language.

Leaving the woods, I enter my pasture, a miniature version of the Serengeti Plain, another mode of the Great Garden. Here, wild and domestic life mingles even more intensely than in the vegetable rows and orchard. I once sowed "improved" grasses and clovers here, believing the universities, which told me these improvements would be better for my cows and sheep than the herbage that nature grew. Nature laughed at such pride and sowed more enduring plants. In almost every case, the wilder ones have proved better for the livestock than the university-improved ones, not to mention for the birds and insects that also live there. Even the "weeds," except some of the more noxious ones introduced from Europe by pioneers who also thought they could improve the native landscape, make good grazing. If I mow occasionally, the pasture takes care of itself.

Meadowlarks sing from fenceposts, bluebirds nest in the houses I have set atop some posts, kingbirds sit on the fence wire between the posts, bobolinks burble and spin up over the fence and into the grass again, barn swallows dart at bugs rising from the grass, field sparrows crouch over nests of eggs at the base of bull thistles. Cowbirds perch on the back of the cow and the sheep, watching for flies. I rake the meadow with my binoculars and gather the whole scene into a spiritual harvest.

I pass into a third realm of the Great Garden: my fields of corn, oats, wheat, and clover hay. Red-winged blackbirds walk the cornrows, stolidly hunting cutworms. I turn over a lump of barn manure that didn't get worked into the soil at planting time and uncover two ground beetles, a species that also feeds on cutworms and wireworms. I lift another manure clump and find two more. The reason for these unworked clumps is that a killdeer had been nesting there at planting time, and I dodged her with tractor and disk. In the wheat plot, a path of trampled stalks leading into the stand tells me that raccoons or groundhogs are probably in the field, digging burrows that the growing grain stalks already hide from view. I scowl, the Mr. MacGregor in me asserting himself.

I pass into a fourth realm of the Great Garden, the grove of trees through which the creek winds. I sit on the bridge I built across the stream, my legs dangling over the side, and gaze into the water tumbling over the rock dam the children built. The sound of water over the stones is spring's best music, next to the meadowlark's song. Along the bank, almost in the water, a wild iris blooms. It appears to have been deliberately planted there, I catch myself thinking, still needing to remind myself that nature was planting flowers long before humans and can do the job just as well.

Suddenly, a fish flies between my dangling legs. It leaps from the water under the bridge in an arc up over the dam into the upper pool. I can't believe my eyes, so I wait. Another one! At least a dozen dance over the dam as I watch. How did these common little shiners and larger suckers (as we call them) learn to leap dams built by children? There are no natural rock dams in our world of mud-bottomed creeks, far from the salmon runs of the wild Mackenzie. And yet, is the "real" wilderness any more spiritually vitalizing than this humdrum remnant left in these Ohio farmlands? If all the land were kept as part of the Great Garden, there would be little need for wilderness parks.

But all land is not kept this way. I walk into a section where, as far as my eye can see, there is nothing but plowed soil. I come here to hunt flint arrowheads and stone hammers left by the Tarzans of another era. I search a while, but the stillness, the eerie emptiness of hundreds of plowed acres stretching into the gathering dusk, overwhelms me. No barns, houses, pastures, woodlands, or fencerows are visible. I have entered a strange planet, one which man has almost succeeded in severing from the full life of nature. Ironically, the men who create these moonscapes for money use the profit to vacation in far-off wilderness areas.

I shiver from some vague fear. A vision of nature decapitated spreads before my mind's eye: a future in which this countryside is slowly but surely turned from its original Great Garden into a desert stretching between lonely roads, a no-man's-land between cities. I see whole townships and counties where a virtually limitless variety of plants, insects, animals, and humans all in their allotted niches once lived—field, pasture, woodland, farmstead, and village—now turned into empty spaces of pulverized, eroding soil producing surpus corn, rootworms, poor-quality food, and an unhealthy society. The Indians left their flints to mark the passing of their culture. I have only a hoe with a shiny handle to mark the end of mine.

I retreat back to country where the Great Garden is still remembered. A wood thrush sings as I approach my tree grove, renewing my hope. The dark vision cannot come to reality, the thrush seems to be telling me, because the continuum of wild nature is even stronger in humans than the continuum of greed. Even the agribusinessmen will understand, once the wilderness areas they escape to are all paved with traffic jams and populated with deanimalized bears eating human garbage. Then everyone will be convinced that the only "escape" is to make all the Earth over into the various realms of the Great Garden.

17

I'm Glad I'm Not a "Real" Farmer

1999

A friend stopped by on his bicycle the other day on his way to town to sell his soybeans. I was taken aback. In our county, it is not seemly for a farmer to pedal a bicycle in public, least of all to town.

"Good thing you're not a real farmer," I quipped. He smiled, understanding that I was referring to a little joke we share. He is also a school teacher, as I am a writer, so by some standards we are not "real" farmers. It is okay for a school teacher to ride a bicycle to the grain elevator to sell soybeans he has stored there, but a real farmer must drive a forty thousand dollar 4x4 to town even to buy an ice cream cone. Our neighborhood is fond of what one of our "real" farmers did a couple of years ago when he was relieved of his driver's license for repeated traffic violations. He ran out of cigarettes while harvesting soybeans, so he drove his father's $200,000 grain harvester to town to get a pack. My neighbor enjoyed my incredulous expression upon seeing him on a bicycle on his way to sell his soybeans. "I hope to have enough money left over so I can buy gas for the pickup when I go to town the next time," he joked sarcastically. Beans were $4.53 that day and rumored to go lower, which meant that he was losing at least fifty cents on every bushel he sold. The only profit was in riding the bicycle, which helps keep him trim. I complained about low farm prices, the warm-up for any farmer talk in 1999, even when the subsequent conversation veers into buying a winter condominium in Florida. He only smiled. "Actually I'm fortunate," he said. "Think how much more money I'd be losing if I were a real farmer with a thousand acres-worth to sell." Our in-joke about real farmers got started when agricultural economists at Ohio State and the University of Kentucky, in reply to criticism that Wendell Berry and I were leveling at them, said that we weren't real farmers, but writers. Later, as I watched Wendell trying to blow the breath into a newborn lamb, I couldn't resist saying: "Good thing you're not a real farmer, Wendell." When, in my own barn, I picked myself up out of the cow shit after Elsie had belted me with her hoof for trying to milk out her swollen back quarter, Carol said it again: "Good thing we're not real farmers, isn't it?" Even my grandson sensed the irony in the situation. As we ranked hay bales in the barn loft where the temperature ranged around 102, he observed: "Sure glad we're not real farmers, Paw-paw."

I try to joke about it, but I do not take kindly to agricultural economists who have not worked on a farm for thirty years, if ever, sitting on their gold-plated salaries and health plans, with pensions that I as a taxpayer pay nearly half of, saying that I'm not a real farmer. The "real" farmers, by the economists' definition, are playing golf in Florida this winter along with the economists, while I am milking my cow, slopping my hogs, haying my sheep, feeding my chickens, and cutting wood for the fire.

What the expert economists mean, of course, is that I don't "make a living" from farming and for that reason my views on ag economics aren't valid. The economists do not seem to realize that they are intimating that *their* views aren't valid either, because *they* certainly don't make a living from farming. As a matter of fact, right now in this wonderful new millennium, hardly any farmer in the Midwest is making a living from farming. As market prices stagnate, industrial grain and livestock producers are living on past investments, government subsidies, and inheritances. Or going broke. If making a living from actual current farm production is the criterion of a real farmer, there ain't any left.

On the other hand, it seems to me that, living at nature's pace on our little farm, I come closer to making my living from farming in a literal sense than "real" farmers. Carol and I raise most of our food including our meat, and some for other family members, keep a garden almost an acre in size, produce half of our home heating fuel from our own wood, derive most of our recreation and satisfaction from our farm, grow corn, oats, hay, and pasture, keep a cow and calf, two hogs, twenty ewes and their lambs, a flock of hens and broilers, and sell a few lambs and eggs. I'm sure I spend more time *living* on our farm than any industrial farmer in our county does. When they are not golfing in Florida or fishing in Canada, they spend a lot of time in the coffee shop or in my office telling me how farming is going down the drain.

In terms of net profit *per acre,* I think Carol and I are making as much money as any of them on our 32 acres, but the total is only a couple thousand dollars, so we're not real farmers. Dividing $2,000 by 32 acres gives me about $60 per acre net return. I can tell you that is more than industrial grain growers who have to rent land—which is most of them—have made the last couple of years. But industrial grain producers have seven ways from Washington, D.C. to manipulate income and expenses for tax savings and can take advantage of subsidies like LDP (loan deficiency payments, which in practice aren't loans at all), plus Freedom to Farm payments (known among farmers as "Freedom To Fail" payments) not available to small part-timers like me. If I were to add to my net profit the amount of money that taxpayers don't pay me for my production by way of direct and hidden subsidies, my financial picture would look even better. But diddling around with a mere $2,000 to $3,000 per year is not real farming in the new millennium. Real farming is having to borrow $200,000 in the spring for operating expenses and hoping to make enough to pay it back in the fall.

In many cases, people who somehow get defined as farmers are much less "real" than Carol and I are. I know a husband/wife team who invested in a pig factory as contract growers. A large company (integrator) pays them so much a pig for finishing 4,000 of them per year in their facility. They do not own the pigs. The integrator also pays for all feed, medicine, and veterinary care. The husband is on the road most of the time with his regular job, and the wife runs the day-to-day hog operation in her spare time—mostly a matter of pushing buttons, she says. Hardly much of that, even. Motors go on automatically to deliver feed and water to the animals, and to regulate temperature and humidity in the pens. Mostly she just checks on the animals and equipment and makes sure there are no problems. She says she can do it all while holding her baby in one arm. If something goes wrong, she calls the company's experts and they take over.

Cleaning out the manure once or twice a year and loading out the fat hogs for market are the only two time-consuming jobs involved, she says. She is a "real" farmer because she and her husband are dealing with "real" money—the confinement building costs them over $100,000. (Needless to say, this pipe-dream kind of "farming" has fallen on hard times as hog prices dropped dramatically because too many "farmers" got into the act.)

In the same way, some industrial grain producers work on their land only at planting and harvest time. Sometimes they even hire custom harvesters. They don't really own their machinery, but lease it by way of "rollover" deals, as other executives do their fleets of cars. (It is hard to figure out who actually does own big farm equipment these days. The rollover deals all are based on the supposition that less well-heeled farmers will buy the equipment as used machinery at an inflated price when the big boys roll it over for new.) Custom sprayers apply all the fertilizers and pesticides. Hired "scouts" watch for disease or insect infestations. Rented semi-trucks haul away the grain. One of the most successful "real" farmers I know hires all his farm work done and spends his time at his computer, attending to what is now a full-time job for the large operator: keeping track of costs, hunting for more tax loopholes, and trying to market his grain profitably by hedging and by puts and calls on the Chicago Board of Trade. He is no more a farmer than is a marketing agent with Continental Grain.

Seems to me that economists have backed their way to the edge of a dangerous precipice when they start defining reality in terms of the amounts of money involved. By their definitions, most of the great works of art and discoveries of science would not be "real" art or science, since they were produced far from the chainsaw of "economics." (Have you noticed in our Tower of Babel world how often it becomes necessary to put words in quotes because they no longer mean what they are supposed to mean?) Name a great artist who produced a masterpiece with one eye on the art market. Name one Einstein who made a major scientific discovery while worrying about his income. Most artists, inventors, and discoverers have made their mark only at the risk of losing money and position.

The same is true of farming: Great discoveries in agriculture are not accomplished by playing the Chicago Board of Trade. What the economists call real farming in fact tends to ignore the best science and art. Any damn fool can grow corn today if the weather cooperates. Technology has taken all the skill out of it. But there is much harm in the trade-off. Diversity of crops gives way to monoculture. Skilled crop rotation with legumes to enrich land and decrease pesticide and fertilizer costs becomes a lost art. Windbreaks are not kept up. Grass waterways are plowed through. Steep hills are planted to annual crops under the mistaken notion that "no till" methods will avoid erosion on such land. The combination of husbandry and agronomy that brings so much true efficiency in soil enrichment and diversity of market opportunities is abandoned. Push-button confinement operations replace pasturing livestock as nature intended, and only an increased use of more and stronger antibiotics keeps confinement production "alive." And still bacteria become resistant to the drugs. Hogs are now genetically developed to survive in confinement if kept on a continuous low-level dosage of drugs, but they can't survive outdoors. A "real" farmer I know bought a bunch of feeder pigs to fatten in the low-cost, old-fashioned way: on bedding in an old barn. Half the pigs he bought came from farms where animals were raised in a natural environment and half from confinement operations. The latter caught pneumonia almost immediately and many of them died despite massive injections of antibiotics.

How can a mega-hog or poultry factory even be considered a farm? More manure is generated by Buckeye Egg in Ohio, which have been repeatedly cited for violations of pollution regulations and is now being sued by the EPA, than from the entire human population of a large city. Yet Buckeye Egg and all large animal factories do not have to install wastewater treatment facilities as towns and cities and other industries must do because these confinement operations are still defined by law as "farms." When Land Grant college economists (most of whom support animal factory technology), say that my 32-acre operation is not a real farm, they are reinforcing the notion that mega-sized confinement operations are real farms simply because lots of money passes through their hands.

A lot of problems pass through their hands too, problems that my "unreal" farm avoids simply by being small. I do not have to use antibiotics or hormones in my animals, for example. My operation does not pollute the air or water. A new study out of the University of North Carolina documents evidence that people who claim various physical ailments from the air ventilated out of large animal factories, a claim once ridiculed by the industry, may be speaking the truth. Most disconcerting of all, Public Health officials in Minnesota have recently released the results of a study that they say proves resistance to antibiotics can be transferred from animal to human (reported in the *New England Journal of Medicine*, May 24, 1999).

I'm grateful that I am not a real farmer and therefore have my own clean vegetables, fruits, chicken, eggs, pork, beef, and fish to eat. I am even more grateful, as I warm my toes under the woodburning stove this winter, that I do not have to battle traffic jams to Florida golf courses or Colorado ski resorts to keep from being bored. Our unfarming life is our recreation and I'll settle for that any day.

18

Going to Market on a Warm Day in November

1999

As I haul a pickup truckload of market lambs to the stockyards, I see November in all directions: acres and acres of new winter wheat, like great green lawns, lying against backdrops of oaks, gold-leaved in the sun. Yellow-brown fields of recently harvested corn alternate with the green swards. Occasionally a deer wanders tentatively through the stubble, confused at the disappearance of its summer-long hideout of standing corn. An Amish boy, stripped to the waist, hoses road grime off a buggy in a barnyard. That's how warm it still is. But the bugs and heat of summer are gone and not particularly missed. The harvest is almost over and not particularly missed either. Barns and cellars bulge with food against the cold siege to come. A quiet calm lingers that no other season can quite match. A nice day in November is the nicest day of all.

I am going to Bucyrus, Ohio, twenty miles from our farm, to sell my lambs. That is the nearest remaining stockyards in the area. In my lifetime, I've watched the yards at Upper Sandusky, Carey, Arlington, Kenton, and Marion close, all succumbing to the hungry giant monopoly that is driving democracy out of agriculture. I am convinced that an economic virus embedded in capitalism causes it to bloat on greed like sheep on frosted alfalfa, and then self-destruct. But I am in a good mood today and do not care to think about troubles. I smile, remembering that at one time, one had to be careful how one used the phrase "going to Bucyrus." Seventy years ago, "going to Bucyrus" was a coy, decoy folk expression in our county for sexual intercourse—especially with a person you were not married to. Bucyrus in the old days was a wide-open railroad town and a mecca of whorehouses and so the phrase was born out of the proper midwesterner's inability to be explicit about such matters. But that is not really the reason for my smile. I am moving the fat lambs to market now in early November so I can put my neighbor's ram in with the ewes to breed them for early April lambing. So, dear me, it is the ewes who are "going to Bucyrus," while the lambs and I are merely going to the stockyards.

I pass a farmer still out planting wheat, too late for northern Ohio. But if the warm spell holds, the seed will sprout and send a tiny shoot to the soil surface, and that is all wheat needs to survive winter unless the weather turns really nasty and there is no snow cover. Planting wheat late reminds me of a story we are fond of in our neighborhood. One unusually wet fall, years ago, we could not get the wheat planted until late October. Willie Masters, who has passed away now, was disking ground late into the darkness of a bone-chilling evening, preparing the seedbed for planting next day. He was all hunched up in a great mackinaw to protect himself from the cold wind. Tractor cabs were as yet unheard of. A neighbor on his way home from his own field stopped in the darkling night to cuss and discuss the lateness of the wheat planting season.

"I was sure you'd be going to the high school football game this evening," the neighbor remarked, making conversation.

"Oh no," said Willie from the depth of his mackinaw. "You'd have to be crazy to be out on a night like this."

Lamb prices are fair this fall, above seventy cents a pound. For the hundredth time I multiply, as I drive along, twenty lambs times 100 pounds per lamb times seventy cents equals $1,400. Not much money in these "good" times, but for those of us accustomed to getting by on $20,000 a year, a nice little dessert on top of the food of daily living, especially since that money will be mostly profit. I have a system, you see, or so I like to tell myself. There is no expensive grain in these lambs, nor expensive hay, nor much machinery or labor cost involved. These lambs have fattened on pasture alone, which is something I never let on at the stockyards. If I did, my lambs might get demoted out of the top of the market because of a myth our cornbelt culture has promulgated for a century: that meat without grain is low quality. To get my lambs sleek enough on grass alone to pass for grain-fed is not easy to do, but I'm learning. I have divided my little fields into paddocks of about one and a half acres each, so that I can rotate the sheep from one to the other, providing them with fresh grass all the time. That means that the lambs not only get nutritious roughage every day, but also get plenty of milk from their mothers. I let the lambs and ewes run together until market day. By that time, the lambs are so big that when they nurse they literally lift the poor ewes off the ground.

The sheep do the harvesting and provide the fertilizer. The only equipment I need is a tractor with a front-end loader, a mower and a rake for making haystacks, plus a sixty-dollar, manual broadcast planter for reseeding. My tractor is thirty-two years old, the mower forty-five years old, and the rake looks like it may have been bought new by Adam and Eve.

But in dry summers, intensive rotational grazing, as it is called, may not provide enough lush pasture no matter how many paddocks I have. Then I have to buy hay or feed grain and there goes the profit. But I'm learning ways to accommodate drouth. Ladino clover, red clover, and alfalfa will be lush and palatable even in the driest August if I can "save" it for that time. If I don't need it for grazing, I make hay from it for winter and drouth emergencies. These legumes will also provide good grazing in November, December, and even early January in a mild winter, if the second or third cutting is saved for winter pasture. Then, in March, a wild grass that came in and took over the paddocks by the creek all by itself will start greening up to provide a bit of pasture in this most difficult of all seasons for northern grazing. By the first of April, the bluegrass and white clover paddocks are ready for grazing. I always plant one paddock to corn and one to oats in rotation with the temporary pasture paddocks (hillsides stay in permanent bluegrass/white clover pasture). The sheep go into the corn paddock after harvest in October to clean up the fodder and the occasional nubbins that escaped harvesting. I usually make hay out of the oats, put in a stack with the front-end loader, and the sheep self-feed from it in winter.

Nearly every day, I stand on the hilltop overlooking the various paddocks and pretend I'm Napoleon strategically planning my "troop movements" from one "battlefield" to another. I get intense enjoyment out of watching animals out on pasture as nature intended them to be, and also out of trying to use rotational grazing so adeptly that eventually the sheep can graze year-round or at least for ten months. But a bigger intellectual pleasure comes from realizing that if I can make this system work on 12 acres, it will work even better on one, two, or three hundred, and provide a young, frugal family with a living.

Bob Evans, the farmer who turned a country sausage business into a restaurant fortune, tells me he has already accomplished year-round grazing with cattle in southern Ohio. Even snow doesn't stop his beef cows, he says. They just paw down through it to get the fescue grass. He has also learned that honeysuckle, which threatens to overtake fencerows in southern Ohio altogether, makes a fairly nutritious emergency pasture if the snow does get too deep for grass. "Year-round grazing could save the family farmer in these hills," he likes to say. "It is just plain stupid that the cattle industry is centered out in the dry West. We can grow grass much better here, and produce animal products efficiently and ecologically. We don't need expensive farm machinery and fuel in agriculture to produce meat and milk." Then he chuckles. "We could dynamite all those stupid silos, too."

The downside of my "system" is that I have 120 pounds of nice wool in the barn that I can't sell, can't even give away. The government in its beneficence has taken away the subsidy shepherds relied on when selling their wool, while foreign wool pours into the country and drives down the price to nothing. Meanwhile literally billions of subsidy dollars go to rich industrial grain and animal factory operators and investors. But I try not to complain about this injustice. The last thing I want is government help.

A friend and sheep-shearing shepherd in North Carolina, Brian Knopp, says that wool makes excellent garden mulch. That's how he is using his. "After rain falls on it a few times, it turns snowy white and rather pretty," he says, his voice hard with irony. "At the end of the season you can pick it up and use it next year."

The stockyards are mayhem, as usual. Men shout, cows bellow, sheep bleat, pigs squeal, trucks roar, auctioneers drone. I pull into the long line of trucks inching up to unload. Another, longer line of trucks jammed with hogs circles around to another dock, and in yet a third line, trucks full of bawling cattle snake to a loading dock at the rear of the stockyards. It is exhilarating to be part of this fanfare, all these loads of cattle, hogs, and sheep streaming in, representing the work of so many farmers and the nourishment of so many consumers. Here capitalism works the way it is supposed to work but seldom does. No matter whether farmers haul in three semi-loads of animals, or one little pickup load, like me, they get treated the same. The competing buyers don't care if farmers run big operations or small; are black, yellow, red, or white; Catholic, Protestant, Jew, or Muslim; Democrat, Republican, or Libertarian; Cleveland Indians fan or Cincinnati Reds fan. Whoever brings in the highest quality animals gets the best price. And the stockyard middlemen will do their best to make sure there is no collusion among the buyers because they get paid a percentage of the gross.

I open the back gate of my stock racks and push the lambs until they jump down off the truck into an aisle-way. I help the handlers shunt them into a pen. A handler takes my name and address, notes on a sales slip that I have ten lambs and one old ewe in this bunch (I'd brought in ten earlier in the morning). The animals are marked as mine and eventually they will be grouped with other lambs according to size and quality. I follow the procedure through the maze of dusty, noisy pens and aisles. My lambs bleat in fear and confusion. They look at me, their shepherd, for rescue. Their eyes are the eyes of my father on his deathbed: why, why, why? All I can answer, as I did to him, is that my turn will come. I ponder, as I often do, about becoming a vegetarian.

Handlers check ram lambs for testicles. If ram lambs have not been properly castrated, the owner gets docked, though there is much dispute among us contrary farmers over whether meat from uncastrated young rams has less quality. I know at least one shepherd who doesn't castrate and sells his lambs on a private, upscale market that demands highest quality.

If the lambs still have tails, and an unreasonable amount of manure clings to their rears as a result, the owner gets docked for that too. I don't cut tails off market lambs anymore because their rears stay clean enough on pasture and without grain feeding. If one or two do get too dirty, usually from eating lush fall regrowth of ryegrass, I cut off the balled manure clinging to the wool with the sheep shears. In Ohio, those balls of hair and manure are often referred to as "dangleberries."

I turn away from my bleating lambs. If I look any longer, I may take them home again. I go to the auction arena, where the buyers slouch in semi-circled rows of seats above the show ring. Lot after lot of animals is paraded through the ring and the buyers bid to the chant of the auctioneer. I grow drowsy under the hypnotic drone of the auctioneer's voice. The consumer who thinks she's a sharp shopper when she buys $600 dresses and then "saves" by buying cheap meat ought to sit here and observe for a while. Most of the animals are healthy and fair to middlin' in condition. A few are sleek and top-notch. But some are scrawny, skinny, old, even emaciated, their sides encrusted with manure. Which do you think sell the highest? Once, in a similar auction barn at Kidron, Ohio, I noticed a man in a suit up in the corner, buying all the worst-looking stuff at giveaway prices. I asked him why anyone would want such sorry-looking cattle at any price. "It goes for processed meats like cold cuts and sausages," he said cooly. "Don't worry, it gets the living hell cooked out of it and it couldn't possibly hurt anybody."

I go up to the luncheonette above and right off the auction arena, where I yield to the temptation of a huge pork sandwich (so much for vegetarianism). These are sandwiches for people who have to worry more about getting shaky in the knees halfway through a physically exhausting afternoon than about cholesterol. Nor is there any place in Ohio that I know, not the fanciest restaurant that I have eaten in, that serves "boughtened" pie as good as those baked here. I listen to the farmers complain about prices being too low, and buyers complain about meat quality being too low. All bitching is done in fairly good humor. The market goes up, the market goes down. Some farmers bring in good stuff, some don't. It is the way of the world and the meat industry knows it. Supply and demand can be denied but it always comes back to bite or kiss you. The meringue on my lemon pie is over two inches high, the warm lemon layer over an inch thick, the crust soft and crumbly. There is never any trouble selling these pies.

I ask buyers and farmers chowing down their sandwiches and pie what will happen if huge agribusiness firms monopolize the meat industry, as seems to be happening. It is common knowledge now that about six international companies control well over half the food business. From "seed to shelf," brags Cargill, which itself is apparently headed for a takeover. From "dirt to dining table," DuPont boasts, as it buys up smaller companies almost as fast as Monsanto does.

"Well, it will be the end of *this* little stockyard, you can bet on that," one buyer says. The others nod. Silence. Then a farmer brings up the currently favorite stockyard story. I pretend not to know it so I can enjoy hearing it told again.

"You could hardly give pigs away for awhile last fall," he says, drawing himself up as if preparing to deliver an oration. "And I do not exaggerate. All those nonfarmers with more money than brains invested in the integrated hog business and integrated the market right the hell back to Depression days. So into the stockyards pulls one of those high and mighty integrators with a big, shiny semi full of feeder pigs. His plan had been, of course, to have his hog-house janitors feed them out for so much a head. But when the price collapsed, he wanted to unload them before he lost any more money. Understand that little stockyards like this one are the very kind of local market that the integrators' private market agreements with packers are driving out of business. So he runs some of his 300 pigs into the show ring. Not a bid. He is looking real desperate by now. Finally, one buyer says he'll take half of them for six dollars a head. [In a normal year, feeder pigs average thirty dollars each.] No more bids. The big shot begs the buyer to take them all. No way. Finally the integrator says, 'What if I *give* you the other 150?' It is so quiet you can hear the auctioneer sucking air. The buyer thinks on that awhile, probably wondering if he could get the integrator to pay him to take the pigs. But finally he nods. So the integrator sold the lot of them for three dollars a head."

Another farmer, trying not to sound gleeful, adds: "And that kind of operation isn't like us little guys, you understand. His costs, labor and such, are out of pocket. He's got to pay up where we can tighten our belts and eat beans all winter."

I go back to the arena to watch the lambs sell. I'm vain enough to yearn to hit the top of the market at least with one lamb. My name would go on the weekly stockyard report and I could brag that a little podunk cottage farmer beat the pros. And then I could make myself even more obnoxious by pointing out that I did it with lambs that had not been fed any grain.

But like other years, I miss the top by a little. My best lamb, at one hundred ten pounds, brings seventy-one cents a pound, two cents under the top. Not Mark McGwire, but Sammy Sosa and that's okay by me. All my lambs sell in the high sixty-cent range except a puny lamb from a set of triplets, which means I know I'm making a little money. This year, the experts say that forty-eight cents a pound is break-even, but my break-even is considerably less because of the way I raise sheep. I get a check for $1,285.

I ruminate on the way home. Most farmers scorn sheep as profitless. We shepherds scorn right along with them, in public. The last thing we need is for hog farmers to switch to sheep. But actually, sheep can be quite profitable. If wool were normally priced, I'd have another $200. If I were a wiser marketer, I might have hit the eighty-cent market that prevailed some weeks. If I were a better shepherd, I would have almost twice the number of lambs to sell from the same acreage. If.

Maybe I can raise fifty lambs on my little place without appreciably increasing costs. Hmmmm. What if I rented some more land and raised one hundred? Two hundred? I once suggested that to my neighbor, Al Kin, a nationally known and respected breeder of purebred Corriedales.He shook his head sadly. "That's always the way it is. A small-scale producer will be doing just fine, and then he gets greedy."

And that's why, in a free economy, there will always be enough to eat and the food prices never get too high. Farmers can't resist overproducing the market and we are very good at it. But what will happen in a monopoly economy, where food prices would be controlled by six CEOs making a million dollars each and living far from the smell of manure and the sweat of the laboring man? Some livestock organizations maintain that business and government have already become so entwined that the antimonopoly laws are a farce.

All I know for sure is that small farmers have no power and never have in the world of entrenched wealth. We live by our wits and run our own underground economy. I can only hope that sheep farming never becomes profitable enough to tempt the monopolists. If it does, I'll switch to snapping turtles. Best-tasting meat in the world, but no one has figured out how to make any business profit from it yet.

19

Looking for a Midwestern Culture

1997

Folklorists at Indiana University (of which I was one), along with a group of Midwestern magazine editors and writers (of which I was also one), used to spend inordinate amounts of time trying to identify and document a Midwestern regional culture-character different from, say, Downeast Yankees, Suthin' belles, Mississippi rednecks, New Yawk guys and dolls, West Coast surfers, and Arkansaw stumpjumpers. It would have suited our literary and folkloristic agendas so neatly if we could have concluded that there really were Hoosiers and Buckeyes and Minnahsooohtah Swedes out there who could be stirred up in a literary doughmixer and baked into a Midwestern cultural apple pie. But as much as we tried, twisting anthropological and sociological evidence until it cried for mercy, we could not in good faith articulate a general Midwestern character that would stand scrutiny as something unique from other regions. In the process of trying, we also realized that New England Yankees, Southern belles, and Arkansas stumpjumpers were no different than Iowa Yankees, Cincinnati belles, or Kentucky stumpjumpers migrated to southern Illinois.

And no one is sorrier about that than I am—or more guilty of playing the game. But I confess my sins. I now know that Kentucky fried chicken is not one gram of crunchy crust better than Ohio fried chicken, especially when fried in Ohio by a Kentuckian, such as my wife. If true cultural differences in various geographical parts of the United States ever existed, like the pockets of highlanders iso-lated in the Appalachians who were still singing Elizabethan folksongs in the early part of this century, it just ain't so anymore. Or is so only at the most fleeting, casual glance. Electronic media have taken care of that. Sometimes on the radio, you can still tell where the station you are tuned to originates by the sounds and patterns of speech that the announcers use, but only rarely. Education has taught us that we should all sound the same. And so it has come to pass. We talk alike, we think alike, we travel alike, and go to places that look alike. Or you can choose your "culture" in any of the alike cities of the nation. Do you want to eat German, French, Chinese, or oil-cloth-farm-kitchen? In every Midwestern town over 30,000 you can have your pick. In every town under 30,000, you have fifty different fast food places to choose from, just like the fifty in every other town. As Holiday Inn ads put it a couple of years ago: "No surprises."

Glossing particular, unique individuals into a common image or mold avoids the difficulty of thinking intelligently about them. As author Scott Russell Sanders, an Ohioan who also lived in Tennessee and is now a professor at Indiana University, says in his book *Writing from the Center* (the best discussion of Midwestern culture or lack thereof that I have read recently):

> The people I met [in college in Rhode Island] . . . claimed they could hear the rustle of cornstalks in my voice, could see the roll of glacial plains in my walk . . . I doubt my origins showed so plainly; but my fellow students on hearing where I was from, saw what they expected to see. Who else but a Midwesterner, they asked me, would refuse to drink alcohol, smoke dope, or sleep with a willing girl, refuse even to swear? I was considered a bit odd back home as well, I answered, but they were not listening.

Not to mention that lots of people grow corn on the East Coast, on "rolling plains" that might produce a similar gait, if indeed that were something influenced by terrain.

I am as pure a native of the Midwest as can be found, which is to say that I am most impure. My parents and grandparents on both sides of the family were cornbelt farmers going back to the Native American era, who also exhibited no markedly unique Midwestern culture. On my father's side, these farm "folk" came from Maryland by way of the British Isles. They married Germans until they were almost as German as the farmers on my mother's side. My maternal great-grandfather walked to this county from Columbus, sixty miles, leading two cows. I live on a part of his original landholdings. My great-great-grandfather on my father's side came a generation earlier, but he was smart enough to leave the cows in Maryland. I grew up on a small, rather hardscrabble farm, but it was a *happy* place, not the mean life that Hamlin Garland and Willa Cather et al. described. Upon reading their novels and others of the genre of so-called Realism, which forever cemented a grossly lopsided image of the Midwestern farmer in the minds of urbanites, I felt insulted and angry. In my ears echoed the words of both grandfathers and my parents, wondering aloud how anyone would be fool enough to want to live in a city.

Then along came the soap opera strains of *The Bridges of Madison County*. Almost immediately, the media hailed the book as an awakening of interest in the Midwest as a region, but book publishers tell me that the book was first a success and established its momentum in New York City. It was a good enough story, well-told, but seated in the old Garlandish image of farmers as living a culturally mean and pinched life in contrast to enlightened urbane society. This suited the abstractionists just fine, since they could then imagine not only how different rural Iowa must be from their centers of urbanity, but also how it was culturally inferior. They had another place to travel to, to see something "regional."

One of my sisters, a *real* Midwestern woman, would tell you that farm wives hate covered bridges; they can't get their mammoth modern equipment through the blamed things. And farm women in the Midwest are not sitting around isolated on farms waiting for white knights from some far-off heavenly city to rescue them. They go (quite frequently) to the city when they want a change of pace. *Bridges* was not about the Midwest. It was about two people fantasizing about what we all fantasize about.

I will admit that when I made the dumb mistake of heading East to seek my fortune in the big city of Philadelphia, I did think that I would come into the embrace of a society that generally rejected the autocratic and hypocritical "Babbitt" Christianity I had known, not to mention the narrow-minded, apathetic, smug self-righteousness about lifestyle that I saw (and shared) in home territory. Guess what? While I met a few of what I considered then to be enlightened people (in about the same proportion as they existed in my corner of the cornbelt), the vast majority of city dwellers in Philadelphia were culturally just like the people in rural Ohio: religiously autocratic and arrogant, poorly read, intolerant in their political and occupational self-righteousness but with the saving grace of being marvelously humorous about it all just like the folks back home. In fact, no more provincial people have I ever met than the typical Philadelphians or New Yorkers, whether they wore blue collars or white. My mother read *The New Yorker* faithfully on her farm in Ohio. I wonder how many people in New York read *The Ohio Farmer*?

The awakening of the notion of a common Midwestern culture is not coming from cultural roots, but from people who have something they want to sell or say, especially media types looking for reasons to start new magazines, radio stations, and TV programs. At a time when economics has driven so many people to the city, and then ripped apart traditional institutions such as families and local schools, they believe that some soft, foam-rubber abstraction like "Midwestern values" will sell. They think they feel an almost palpable nationwide yearning to live in small towns surrounded by cornfields, where kids can walk to the store alone without fear, where people really do still sit on front porches and count fireflies, and are not afraid that they will be accused of molestation if they hug the child next door. But if people are so yearning, why don't they do what those of us who live in little villages do: Come on back and work at nature's pace for half the salary they get now?

The truth is, Americans as a whole only pretend they are yearning for small town or rural life. That is why so many attempts to sell something clothed in "Midwestern values," or any other kind of "regional" baloney, fails, including running for public office on that plank. A good way to see through this hypocrisy is to try to get a doctor to move to a small Midwestern village, as we are doing. It is, in fact, the emphasis on this regional schlock charm and character that scares thinking people away from small towns and rural areas. If magazines and television and newspapers made them aware how *un*-uniform we "regional" "folk" are, how full of surprises are our towns and rural areas, how different we are one from another, just like everywhere else, then more people would be inclined to come here. Geography has nothing to do with our differences and similarities. The magazine or television station that uses statistical demographics as a basis for circulation rather than intelligence and wit is doomed to the region of Ho-Hum.

It would be easy enough to explain the seeming fascination with a Midwestern culture by pointing out that local color in all other areas of the country has been exploited until it is all off-white. But there is much more to it. Americans, particularly Americans working in the media, are rebelling against the mass culture of consumerism. Mr. Sanders says it well in his book cited above:

> America is still a manifold and textured land, nor could any human force ever wear it down to utter uniformity; but each year it is rubbed a bit smoother by technology and commerce, by the media, and by our feverish mobility. The more we gad about, chasing dollars and dreams, the less of an imprint we bear from any particular place, and the less likely we are to know in depth the place where we happen to be. . . .

The dilemma for the media "folk" is that they have supported the uniformity they rail against, if not actually orchestrated it, because their television stations, magazines, and newspapers need a mass culture to achieve the kind of circulation that allows them to enjoy big salaries. But trying to appeal to the largest common denominator inevitably means producing artistic or literary pap that hardly anyone with a farthing of intelligence reads or watches. It is like replacing the piquant tastes of a properly spiced and seasoned seven-bean soup with a bowl of hasty pudding. No surprises. No one particular likes it, but no one particularly dislikes it either—in commercial journalism, the perfect kind of stuff to fill up space between the ads.

The movie *Fargo* was touted as an example of the new consciousness of a unique Midwestern culture. *Fargo* was exactly not that. I think it was a good movie, but it was not really about Fargo and certainly not about the "culture" of Fargo. It was about how outwardly ordinary people can respond to a terrible situation with forgiving love and intelligence or totally depraved cunning. True, the movie used the regional dialect and woebegone hesitancy that we associate with Scandinavians on the northern plains—and did it perfectly. I know because I lived in rural Minnesota for four years. But change the dialect to Maine and punctuate the sentences with a laconic "Ayuh," and you have northern New England. Garrison Keillor's Lake Wobegon in Minnesota has been popular among all of us who live in rural America because we all recognize our own villages in it. Lake Wobegon is not unique. Like television's Mayberry, it is part of the same human community. These "places" are not regional; they are a *universal place*.

Unfortunately, attempting to isolate a regional or local culture, even where there is one, has the opposite of the effect intended. Rather than honoring particular places or particular people, image-making merely caricatures them. Once made into a caricature, the local or regional "character" is marginalized and minimalized as a mainstream personage. Calling Grant Wood a "regional" painter because he mostly painted scenes in Iowa where he began life as a farm boy is how the urbane art community of New York and Paris keeps him in his "place." But Wood lived awhile in Paris too. He came back to Iowa, observing: "I realized that the best ideas I ever had came to me while milking a cow." I have a hunch that Virgil in ancient Rome, or Tolstoy or Chekhov in peasant Russia, could have said the same thing. Antonin Dvorak lived in Iowa awhile and then wrote the immortal New World Symphony. By the thinking of the image-makers, wouldn't Dvorak's work be "regional" too?

Art is forever at odds with generalities, committee meetings, reader profiles, and "demographics" that are drawn with an eye only to making money. "Mass culture" is an oxymoron, a concept good only for making war and dictatorships. The world everywhere, even the so-called dull Midwest, is full of surprises constantly confounding the image-makers and the money-makers. May it always be so.

20

The Folly of Trying to Repress the Agrarian Impulse

—Thoughts while watching weeds push up through a shopping center parking lot.

1999

In the spring of 1999, for the first time ever, urban people who had no money to gain not only marched and demonstrated to save farmland, but raised a cool million dollars to make it happen. Although outside the local area the event passed without fanfare, it may in retrospect come to be hailed as a turning point in this cycle of history, the first indication that a largely urban society is beginning to understand that life is impossible without agriculture. The place was Yellow Springs, Ohio; the land kept out of the hands of commercial developers was Whitehall Farm, which lay invitingly right next to the town limits, in the center of a growing metropolitan area bounded by Dayton, Columbus, Springfield, and Xenia. It was prime development land, all in one, big, luscious 940-acre chunk.

Yellow Springs (population 4,000) is not your usual Midwestern village. Just getting a MacDonald's restaurant in town twenty years ago almost caused a local civil war. Officials of most rural Ohio villages favor "growth," any growth, no questions asked, bowing in adoration before the myth that growth is always good for the local economy, except (cross yourself) growth in whorehouses and bars. But in Yellow Springs, even the village leaders were suspicious of turning Whitehall Farm into another sprawl of houses and business places. A woman who was organizing a little army of demonstrators on the day of the sale reflected the crowd's mood: "When we drive outside the town limits, we want to see farmland, not a jam of strip malls and trash stores and houses that all look the same." Her helper added: "Look, we're not stupid. We know that as long as population increases, there's going to be urban sprawl. But why can't there be places where people by mutual consent are willing to limit family size and spending to stop growth? If this is a free country, why can't we foster a few no-growth communities?"

And so the lines of battle were drawn, one side insisting that commercial growth was a necessary and unavoidable fact of life; the other insisting that some places needed to be removed from this necessity, especially since farmland acreage was continuing to shrink at an alarming rate. (Ohio loses seventy-seven acres of farmland every day to other uses.) Money always settles conflicts about growth, and since commercial developers usually hold the high hand in this regard, they win.

Yellow Springs understood that fact, and so the people did not just stand around talking piously about the issue and doing nothing, but got to work to earn the money. It was a genuine grassroots people project. There were some big donations, true, but also a whole lot of small ones, the money raised dollar by dollar with bake sales and car washes and local festivals and door-to-door solicitations. Led by local lawyers, governmental office holders, and retired professors at Antioch College, which is at the heart of Yellow Springs, plus many just plain angry and fearful citizens, the town established a Land Trust to act as a legal entity to buy the farm and place it under conservation easements in perpetuity. That meant the land could be resold only for farming or open space. It also meant that the Trust would have to raise enough money to cover the difference between what developers would bid for it at auction and its farmland value, so it could be sold back to farmers at the latter price.

Few businesspeople or experienced observers of farm preservation attempts thought the "Save the Farm" coalition had a ghost of a chance of success. The money needed seemed an impossible amount for a town of 4,000, some of whom would rather have the land developed. They also had to find farmers willing to go along with buying land that would remain farmland forever. Most farmers piously lament developmental pressure on their farms but when they retire, they want to sell it to the highest bidder.

To make a long story short (a detailed account of the land sale can be read in "Showdown At Yellow Springs," by this author, in *LandOwner,* vol. 21, no. 6, April 5, 1999), the people did raise the money. Their lawyers, who also were buying some of the land to keep in farmland for themselves, simply outbid everyone else. The property sold for $3.2 million. The Trust had no trouble finding farmers to buy land under conservation easements for about $2,000 an acre, with the $1.2 million raised by the Land Trust covering the difference. Large cash-grain farmers and small organic market farmers were eager to buy, in fact. So another myth, the one that says farmers won't buy land in conservation easements, bit the dust.

But the showdown at Yellow Springs was very much the

exception to the rule, so far. Whether it heralds a new attitude about farmland on the part of both urban and rural people remains to be seen. In the meantime, government is pretending to want to save farm-land by pointing out the evils of "urban sprawl." Although no one has stumped more vigorously for farmland preservation than I have, much in the current opposition to "urban sprawl" seems to me to be fruitless or deceitful. First of all, how can society take the need for farmland preservation seriously when landowners are being paid millions and millions of dollars to take millions of acres out of production under the Conservation Reserve Program? Or how can people seriously believe those of us who worry about land loss when we are raising huge surpluses right now that drive down farm prices and drive farmers out of business?

Close investigation reveals that "urban sprawl" is the same thing that was called "rural development" twenty years ago, which was not rural development but industrial development in rural areas. Ironically, the same commercial, governmental and educational offices that trumpeted rural development now want to damn it as urban sprawl. An example of this contradictory turnabout is the huge Honda plant that spreads over at least one hundred acres of prime farmland near Marysville, Ohio. When the Japanese were throwing "development" money around like holy water, their sprawling, land-wasteful plant was hailed and hallowed by bank, political party, university, and the Farm Bureau, the mouthpiece of big agribusiness, as great "rural development." A company trying to do that today is accused of "urban sprawl."

Finding factory jobs for a wrecked rural culture is not rural development but rural dismantlement. But that didn't become apparent to mega-agribusiness until rural dismantlement began to compete with them, not just small family farmers. Then rural dismantlement became urban sprawl and had to be contained to preserve the monopoly that industrial grain and meat factory investors enjoyed in acquiring more land. It was all right when mega-farmers could run up the bid on land beyond the reach of small farmers, but it is suddenly not all right now for commercial developers and private citizens seeking a homestead in the country to bid up the price of land beyond the reach of mega-farmers.

But if, with small farmers gone, farmland is to be preserved by blocking off huge chucks of land for the exclusive use of industrial food producers, as some opponents of urban sprawl proposals suggest, how is that farmland preservation? I think it is factory farm monopoly preservation. Industrial food consolidations, aiming to control food from seed to shelf, don't want competition in land acquisition and are taking advantage of naïve advocates of farmland preservation to gain even more monopoly. Moreover, they are themselves "urban sprawl." In fact, the expansion of large industrial grain operations and huge animal factories displacing farms and destroying rural communities appears to me as a kind of "urban sprawl" that is worse than strip malls and housing tract development. Huge hog factories, financed by urban money (even mutual funds) and operated by paid managers and ill-paid immigrant workers, are just more industrial urban development in the country.

The meetings held to discuss urban sprawl, which was last winter's agricultural college craze, revealed how divided and hypocritical society stands on the subject. The group most legitimately opposed to urban sprawl (as they were to "rural development") are family farmers who want to keep their land in their families forever, if they can, but fear that commercial development will bring property-tax escalation that will force them to sell. Another group, large-scale animal factory operators, oppose urban sprawl, other than their own of course, not so much for tax reasons (since they get property- and personal-property tax breaks not possible to small farmers), but because they know newcomers will make trouble for their polluting facilities. The third and most conspicuous opposition group, the ones that I say agribusiness is manipulating for its own ends, are ex-urbanites who have already moved to the country and now wish to keep other people from doing the same.

Most older farmers, land investors, and developers support urban sprawl just as they supported rural development, for obvious financial reasons. Farmers want to be able to sell out at the highest possible price when they retire, as I have said; developers want to buy land that is cheaper than property in town; and land speculators, seeking a balance to the nervous jitters of the stock market, want to be free to wheel and deal in land while in the meantime raking in "Freedom To Farm" and CRP payments. *LandOwner* magazine reported last year that a wealthy farm couple from Illinois stopped by the big Plummer land auction near Johnson, Kansas, while on vacation. On the spur of the moment, they bought a 160-acre tract out of the estate for an "investment." Writes editor Jerry Carlson: "Their 'hobby' is collecting extraordinary farmland buys. They felt the quarter-section they acquired for $65,000 will be a good investment. Their ten years of CRP payments will return $56,360."

The debate over urban sprawl seems to me to exhibit little realistic thinking on either side. No discussion has taken place of slowing down population growth, which precludes any rational plans for saving farmland in the long term unless people are forced to live in tenement housing and highrise apartments and live on algae. No discussion has ensued over the mad continuation of extending new highways over rich farmland. Little attention has been paid to the unavoidable fact that growing towns and villages must extend corporation lines into the countryside because they don't have any inner-city space to use more efficiently. No proposal has been made to stop government from acquiring farmland whenever it pleases for prisons, parks, research sites, hazardous and radioactive waste dumps, airports, or utility rights-of-way. Indeed, when Ohio State received, mostly as a gift, the 1,500-acre Firestone estate near Akron two years ago, it had a perfect opportunity to practice what it was trying to preach at the meetings and sell that land (or lease it long term) to young farm families at a reasonable price. Instead, the university sold the property for $12 million to a big-time developer who plans to subdivide it into acre and a half lots, which opponents of urban sprawl consider a "wasteful" use of land.

The best suggestion anyone has come up with so far is for the state to buy up developmental rights to farms in exchange for the farm owners agreeing to keep the land in farming permanently, much like the citizens of Yellow Springs did. But as Gene Krebs, Ohio state legislator, who favors farmland preservation (and one of the few bonafide farmers on the Ohio Task Force for Farmland Preservation) pointed out, such legislation has only very limited application. "The math just doesn't add up," he says. At $1,700 per acre for developmental rights, the average nationwide, even a program with a budget of $260 million a year, as has been suggested for Ohio, could buy development rights to only "an infinitesimally small amount of farmland."

The most concerted focus on the part of farmland perservationists seems to be to stop new homes from "sprawling" into the countryside. I find this unconscionable, because most of this effort comes from people who have already moved to the country and wish to stop others from doing so. More importantly, a very significant number of people moving into the country do so to farm on a small scale, or to raise huge gardens or specialty livestock, or to grow and perpetuate groves of trees. Often they commit their small acreages to "wildlife sanctuaries," increasing the diversity of wild plants and animals. This is an extremely beneficial type of "urban sprawl." In our county, many such homesteads are acquired and built in the few remaining rural tree groves, saving them from the bulldozers of industrial corn farmers gone berserk.

It seems to me that development plans that would "block off" acreages where only industrial farming could be engaged in would be playing right into the hands of the giant agribusiness firms and their drive to monopolize the food business. Such segregation would also be repeating the mistakes that cities have made. In *Cities of Difference* (Guildford Press, 1998), Jane Jacobs, noted writer on the subject of how cities grow and develop, argues that human neighborhoods should be a little "messy," that is, they should avoid society's hellbent preoccupation with segregating humans and human activities into neat, everything-in-its-proper-place kind of development. That leads to stagnation and decline, she maintains. All you have to do is drive through Detroit, as I did recently, to see what she means. Like so many other large cities that became domains of megalithic big business, widespread parts of this metropolis look as if they have been bombed and the process of clean-up were still in progress. What has been left are decaying hulks of buildings erected during the heyday of the auto industry when the reigning philosophy was to zone everything into neat and proper places and use the automobile to move people from one neat and proper place to another.

The suburbs continued the policy. No "mess." A gas station/repair garage close to houses where it might be particularly useful as they are in villages? Heaven forbid. Chickens in the backyard and vegetables on front lawns? How trashy. Backyard composters to turn food waste, leaves, and grass clippings into valuable fertilizer? Might draw rats. Little factories mingled in with the homes? Might make a little noise to disturb His Suburban Highness on Saturday morning, although with the roar of lawnmowers, who could hear anything else? Factories, without a neighborhood of active, influential citizens, workers, and owners *nearby* to enforce clean operation, become more and more careless about air quality control, until their smokestacks belch not just on the ghetto people who have to suffer their proximity, but the richer people in the suburbs. The rich people move farther away. As conditions become more inhuman, and as the economic gods demand yet more consolidation and expansion of production and sales, people flee once-nice neighborhoods for new ones on the (current) edge of the city, and the whole round of growth and decay repeats itself. Churches, always the last root of a community to die, tell the story. In the last decade over fifty of them have been abandoned in Detroit.

The same "development" now affects rural areas. Side by side, you can find yesterday's animal factories, now obsolete, with today's animal factories on their way to obsolescence. The schools and churches stand abandoned because the people who can afford to do so move away, hoping to get away from the stench of huge animal factories, from the endless strings of trucks rumbling in and out of the factories, tearing up country roads. Poor people must stay behind. They work in the animal factories and stay poor. The ghetto moves to the country. Emptied rural areas and emptied city centers of industrial exploitation start to look uncannily similar.

Ironically, it is urban people who are showing that this situation can't last, that the agrarian impulse can't be repressed. They are the ones moving to the countryside in unprecedented numbers, at least here in industrialized Ohio, to establish new kinds of farms in spite of and in defiance of the misplaced fervor to stop "urban sprawl." Andrew Stevens, editor of the *American Small Farm* magazine, wrote in his April 1997 editorial urging a positive attitude toward so-called urban sprawl:

> As farmers, you should be looking at your new neighbors as friends and potential customers for your products. Farms that have flourished in rural/urban fringe areas have done so largely because they cater to the preferences of nearby residents. They tend to be more specialized in producing high value crops, sell more products directly to consumers, are smaller, and make more efficent use of resources.

But urban people are also bringing agrarianism back to the cities. Developers build subdivisions that look and function like yesterday's villages or neighborhoods. Gardens and home businesses are planned into the landscape, as are nearby retail and service shops. Some communities even utter the almost forbidden words, "neighborhood schools" again. New neighborhood houses of worship in the ghettoes, small and humble and unassuming, return in the shadow of the abandoned cathedral-like churches. A surge of market gardening and farmers' markets recalls those years not so long ago when thousands of tiny truck farms, using horse manure for compost in their hotbeds and coldframes, supplied their cities with vegetables and fruits nearly year-round. The term "urban farming" turns out not to be an oxymoron. Chicago is even encouraging animal husbandry as part of its urban farming projects. In the heart of Cleveland, in the shadow of skyscrapers, horses plow garden plots. And with the returning agrarian spirit comes its wonderful offspring, agrarian ingenuity. Kathleen O'Neill smiles as she tells me the story of her "Manure Maneuvers." She works with "Cultivating Our Community," the inner city gardening project that borrows the aforementioned horses from Farmpark, a sort of working farm museum in Cleveland. "Cleveland Metroparks Zoo produces and composts more manure than it can use on site," she says. "But we didn't have an efficient way to get the manure to the inner city gardens." She and her fellow workers put their heads together. They contacted the Ohio National Guard, which spends most of its time on weekend maneuvers clogging up the highways with long lines of trucks. Would they consider hauling manure instead? Of course. "Hauling manure fit perfectly into its Civilian/Military Innovative Readiness Training program!" O'Neill says, displaying her impish Irish grin. "Here's an example of what community can do. Several diverse organizations, each meeting its own needs, working together to make the world a green and peaceful place, one garden at a time."

Agrarianism is the best word for it.

I think I hear a faint rustle under the blacktop of shopping center parking lots, under the abandoned animal factories of yesterday and those yet to be abandoned tomorrow. Not only are the weeds pushing up through the cracking pavements, making way for the trees, but the irrepressible agrarian impulse is pushing through too. As long as humans are free to follow their hearts, there is hope.

21

The Wheel of Life Turns Round and Round

—Thoughts while contemplating a woodlot that forty-four years ago was the highest-yielding cornfield in the world.

2000

Three years ago, a nationally renowned agricultural economist made a prediction on the radio that I have a hunch will embarrass him greatly if he lives long enough. (Perhaps all our attempts at predicting the future would embarrass us greatly if we live long enough.) He said that a continuation of larger and larger industrial grain farms and animal factories was "inevitable." It was obvious that inevitable also meant irreversible in his mind. He did not make this statement as his opinion, but as a fact, one that sentimental old agrarian cranks like me had better get used to.

I wonder if he would have made that prediction had he known the history of any particular farm deeply or if he had known that the proprietors of the factory farms he was so arrogantly extolling in 1997 would be standing like bums in a soup line in late 1998, waiting for huge government handouts to keep them financially afloat. In 1999, almost half the income farmers received came not from farming, but from the government.

"Inevitable" is a word that probably ought to be stricken from the language of human behavior. History demonstrates, time and time again, that in agriculture, as in any economic activity, change is the only inevitability. It is just as possible for farming to go from big to small in size as from small to big—to disperse into many units as to consolidate into a few. Nor is the supposed normal progression of land development from wildland to metropolis inevitable and irreversible. I need look no further than right here in the fields of home to see that lesson written on the land. If Wyandot Chief War Pole, after whom the creek that runs through our farm is named, had been a conventional economist, he would have insisted, fifty years before the influx of white settlement, that the continuation of the Wyandots' highly refined combination of farming and hunting was inevitable and the Delawares and Shawnees had better get used to it. But War Pole lived to see sheep ranching become the characteristic agricultural activity here after the Wyandots were tragically shipped off to Kansas in the mid-1800s. War Pole's people and their ancestors had unwittingly prepared for the ranchers by creating, with their annual fire-ring hunts, extensive treeless prairie pastures ideal for sheep. My great-grandfather Charles Rall, coming here from Germany, leading two cows all the way from Columbus, went to work on the R.M. Taylor ranch that spread over much of the farmland I have roamed since childhood. Had there been agricultural economists in those days, I can just hear them saying, with all the pomp and ceremony of their royal offices, that a continuation of huge sheep ranches was our "inevitable" future and hired hands from Germany better get used to it.

But within a generation, money was finding different paths to follow. The sheep ranches rapidly became "obsolete" (in human civilization, obsolete means "something no longer money-profitable enough to compete with something else"), and farmers like my great-grandfather bought up the prairie along with the forests around it and converted it to more-diversified, large livestock farms. So profitable was this kind of farming for awhile that by 1900 Great-grandfather had consolidated some 2,000 acres into his operation, and his was not the largest. Get big or get out, I can hear the economists boasting as if they were responsible: a continuation of large grain and livestock farms was "inevitable." But Great-grandfather's four sons inherited that land, divided it into as many parcels, and eventually distributed it out to their numerous offspring in 160-acre family farms, as I have described earlier. That was the most "efficient" way to apply humanpower for profitability at that time. By then, agricultural economists were on the scene, having found a way to milk the tax rolls for their salaries, and they rallied to that redistribution of land with gusto, declaring its continuation not only inevitable, but a great victory for American free enterprise.

Not quite two generations later, after World War II, "free enterprise" began displacing these little farms with industrial cash-grain operations headed back in size to about the acreage of Great-grandfather's farm. The economists, once more displaying not a whit of historical sense, said, and still say, that the continuation of this megafactory farming is "inevitable."

I can gaze up Warpole Creek from my farm and see the once-forested valley that livestock farming kept in pasture for one hundred years. In that pasture rises a Hopewell Indian mound whose people would have been clearing this land and planting corn five hundred years before it grew back to forest and a thousand years before Great-grandfather (or perhaps the Wyandots) cleared it again. When the sun slants low in the west, I can see Great-uncle Albert's old furrows under the grass of the valley slopes. He tried to grow corn on these hillsides too, until he understood such land was better kept in pasture. He even tried to grow corn on the mound itself! Today this little valley, found not to be accessible enough for the monster machines of industrial grain farming (after a brief flirtation in that direction), is growing back to timber again!

Whose agrarian vision do you want to vote for? Reading what the highly literate Wyandots said about their way of life, I am convinced that they had affected not only the most ecological farming ever done here but were the happiest farmers. Of course, the Hopewells' agrarianism might have been just as pleasant as the Wyandots, for awhile. There is no way to know, because they did not leave a written record of their marvelous integration of farming and commerce and hunting-gathering, an economy that archaeologists say was so successful that it eventually (inevitably?) generated overpopulation and collapsed. But you can bet that right up until their decline began, maybe even after it started, the shamans who passed for economists in the Hopewell villages stood atop their mounds and declared with immense bravado that a continuation of bigger and more elaborate earthworks was "inevitable."

Between the shift from Great-grandfather's large livestock farms to his grandchildren's small family farms, something else happened that further reveals the sham of economic prediction that ignores the history of place. The corn and soybean fields that I can look out upon from my eastward windows today were for a time an airfield! It was known as "Rall Field," naturally enough. The year was 1930. The prophets of inevitability were all talking about how there would be a plane as well as a car in every garage someday.

My kinfolks' airport is remembered not because it violated the conventional theories of historical progression, but because of a humorous story that went with it. On Sundays, planes would fly in from Bucyrus and Marion and other towns in the area and take people for rides. The planes were mostly fragile, homemade affairs guaranteed to supply plenty of weekend excitement. The story goes that the owner-builder of one such plane, possibly not trusting the flimsy crate himself, hired a pilot to fly it from Bucyrus, where he kept it, to Rall Field for an afternoon of rides. Arriving at the field before his plane, the budding airline executive noticed a dead furrow across the upper end of the landing strip. Though only a slight depression in the ground, it might spell disaster if the plane crossed over it during landing. So the first and only air traffic controller Mifflin Township has ever known straddled the worrisome little remnant of bygone agrarianism, and as the plane hove into sight began waving his hands and pointing down to the ground at the source of his consternation. The pilot interpreted the pantomime in just the opposite way it was intended. It seemed strange to him that his boss wanted the plane set down right in front of what looked like a dead furrow, but it seemed obvious, from the increasing ferocity with which he waved, that such was the case. Down he came, as close as he could to the spot his screaming, purple-faced air traffic controller was pointing at. When the plane hit the furrow, it nosed over and crumpled up like a paper accordion, but the rate of speed was so slow that the pilot walked away unhurt.

Rall Field did not last nearly as long as the vision of an inevitable plane in every garage. Great-uncle Albert, applying his astute pencil stub to the daybook he kept handily in his bib overalls, calculated that corn and dairy cows on that land were more profitable than inevitable airplanes, at least for the time being. Today, he might have concluded that a golf course was more inevitable than dairy cows except that other farmers not far away have already reached that conclusion. There are now more golf courses in our county per resident (24,000 population, four golf courses) than possibly anywhere in the nation. This was one "inevitability" entirely missed by the economic prognosticators.

Many farms have strange tales to tell. The first officially recorded 300-bushel corn yield was grown by Lamar Ratcliff on his father's farm in Prentiss County, Mississippi (much to the chagrin of the cornbelt), back in 1955. How well I remember the excitement among us farmers. The farm magazine rhetoric flowed with the promise of mighty things to come. Soon 300 bushel corn would be common. And if Mississippi could do it, by hickory Illinois and Iowa with the help of more fertilizers and chemicals and hybrid vigor and technology, yawn yawn, would soon ring up 400-bushel yields. The word "inevitable" was flung around very loosely on that occasion too.

Today, forty-four years later, the field that grew the first 300-bushel corn is a woodlot again! Furthermore, yields of 300 bushels per acre were not achieved again for twenty years and then only in a dozen or so isolated instances. Ironically, agronomic experimentation indicates that if 300-bushel yields ever do become "inevitable," they will be a product of biointensively managed, raised-bed garden plots, where extremely high yields of almost everything have been achieved in recent years. Imagine how much food the backyards of America alone could produce at such super yields (or even at average yields), putting the lie to the claim by the mega-food companies and their university hirelings that big business is the only and inevitable solution to the future of food production.

We have been here before and know this lie. Way back in 1907, an economist in England, Prince Kropotkin, a forerunner of the new breed of ecological economists today, clearly demonstrated with pages of data how food and manufactured goods could be produced just as abundantly and economically on small, dispersed farms and shops as on the huge bonanza acreages and large inhumane factories that greed was generating at that time. His book, *Fields, Factories, and Workshops,* correctly predicted the demise of the bonanza farms and the "farming out" by large factories of much of their manufacturing work to smaller, more efficient shops. Consolidation is not a synonym for efficiency but only for power.

That the current trend to consolidation in the food business will "inevitably" reverse itself is as justifiable a conclusion as assuming that a half-dozen huge companies will "inevitably" monopolize and control the food supply. All over the "inevitably"-industrialized cash-grain county I wander daily, small homesteads and garden farms are popping up. Some of them are merely urban homes in the country, but a surprising number are younger couples or retirees coming back to reclaim and use some of the land that the economics of power took away from their fathers and grandfathers. When farms go up for sale around here now, they are invariably split up into small parcels of five to forty acres, because there is such a demand for such acreages. Young families with urban jobs can and will bid more for the small parcels than the mega-farmers can afford to pay, a reversal of a thirty-year trend. The mega-farmers fume at the practice of selling farms in small parcels, forgetting how they said, in their day of breezily buying every farm that came up for sale, that this is just "the good ole' American way."

Why this new development in rural life receives little attention from the economists is beyond me. (Not really. The people spearheading this "forward to the land" movement do not want or ask for political or educational help and so of course the bureaucrats must pretend they don't exist to save face.) My own family makes a good example. My parents raised nine children on a typical family farm of the thirties and forties. Eight of us children live rural lives today much like our parents did, two of us heavily into commercial farming but with other sources of income, and the rest of us dividing our time between other careers and small-scale farming, sheep raising, orcharding, tree farming, and very serious subsistent gardening. Of our twenty-nine children, fourteen are married, and of that number, nine live rural lives as we do, and three of the five others tell me they will move forward to the land as soon as they save sufficient money to do so. Of the remaining fifteen still in school or still single and trying to find their way, I am certain that at least half will eventually (inevitably?) take up our rural lifestyle. And we are outdone in this respect by other rural-rooted families in our neighborhood.

We "new" countrysiders express an allegiance to the same agrarian values that our parents and grandparents and great grandparents honored. The only difference is in our way of expressing that allegiance and in the work we pursue to achieve it. We come back to rural life because we want some physical control over our lives. We are rebelling against the economics of power. We want some income from the land but also some from non-farm sources because we understand the folly of trying to make a decent financial income entirely from farming in today's power economy. We want homes where our children can know meaningful work and learn something useful as they grow up. We want an alternative to chemicalized, hormonized, vaccinized, antibiotic-treated, irradiated factory food. We would like to establish home-based businesses when possible so that we do not have to put our children in day care centers as parents who work away from home often must do. We want a different kind of educational environment for our children than what consolidated, power-economics schools provide, private or public. Sometimes we homeschool our children. We want, above all, some home-based security not dependent entirely on power economics. We think the economics of power has run its course in this cycle and is going to hell.

What we are doing, in short, is finding ways, which the farm organizations failed to do because they taught farmers to put all their eggs in one economic basket, to bring back to rural America the life and money that consolidated banking sucked out over two centuries of predatory colonization, a process that consolidated schooling legitimized. We are managing to join the best of urban life with the best of rural life in a new and admirable agrarianism. Steve Zender, editor of one of our local village newspapers, wrote a telltale anecdote in a book he published in 1998 in celebration of rural and village life, *The Big Things in Life Are the Little Things*. One of his reporters, Kate Orians, left a message on his office phone: "I'll bring my story over Sunday evening or afternoon. It just got too late today and the vet was here to take blood samples from the pigs." Zender laughs as he tells the story again. "It is wonderful to live in rural America where we have the best of both worlds," he says. "Futurists say that telecommunications will result in people fleeing the cities and moving to rural areas. It's already happening here and I hope it means that people will be able to provide for their families and do important work for their communities while still having time to garden and take care of pigs."

The most interesting and promising ideas in food production are showing up in this "new" agrarian society. While mainstream factory farming continues to cement itself financially into huge cumbersome operations that lose the flexibility to move quickly to take advantage of changes in consumer-driven markets, small-scale farms are perfecting new/old practices such as deep-bedding systems for hogs that are free of factory farm odor and pollution problems; organic dairies; meat, milk, and egg production that relies on rotational pasture systems, not expensive chemicals or machinery; permanent, raised-bed vegetable gardens where production per square foot is enormously increased with hand labor, not expensive machinery; improved food plants from natural selection of open-pollinated varieties whose seed can be economically saved for the next year's crop; and so forth. The mindset of the new small farmers is not simply traditional. The seminar that drew the biggest crowd at a recent small farm conference in Indiana was on raising freshwater prawns. The thinking of these new farmers is rather un-traditional, too. Chip Planck, a successful, long-standing commercial market gardener near Washington, D.C., used to be a professor of political science.

Even the schools of economics are headed in a different direction than many conventional economists realize. New economic theories have reverence for historical evidence. One of the foremost spokespersons of this different philosophy of economics, John Ikerd at the University of Missouri, recently wrote a paper titled "The Coming Renaissance of Rural America." Discussing change in food production cultures, he asserted the validity of the "universal cycle theory," which *Science* magazine recently included in its list of the top twenty scientific ideas of the twentieth century. According to this theory, any observed trend is in fact just a phase of a cycle. "If we look back over past centuries and around the globe, we can find examples where control of land became concentrated in the hands of a few only to later become dispersed in control among the many," writes Ikerd. "The trend toward fewer and larger farms in the U.S. might be just a phase of a cycle that is nearing its end."

The ending of this cycle, as the ending of any cycle, will be sad, even tragic for those who do not see it coming and so do not change in time. The gigantic, high-investment food factories of today, like the 10, 20, 30, and even 75 thousand-acre bonanza farms at the turn of the twentieth century, do not in fact have the flexibility *now* to respond fast enough to changes in market demand. If just 25 percent of the people in this country decided to become vegetarians, or if 40 percent decided to cut their consumption of meat significantly, which is entirely likely, the factory farm system irrevocably based on meat, corn, and soybean meal is history. Like bonanza farms, like any dinosaur without great adaptability to a new environment, such a system must die out until conditions are right for it again.

But in every other way than the financial welfare of the bonanza crowd, a swing back to a more distributive food production system will have many advantages for the common good, not the least of which would be an end (until the next cycle) of the food monopolies now being created.

The problem I fear is that while we are condemned by the economics of money greed, or changing markets, or weather patterns, to continue the inexorable cycles of small and large, boom and bust, consolidation and dispersal, the power of wealth solidified in the current cycle will try to stop the wheel of history from turning to the next cycle. Expect a vigorous effort on the part of the agribusiness oligarchy, in cahoots with a suppliant government, to prolong the bonanza farm consolidation of today as long as it can. Some of this futility will be justified to prevent a period of chaos in readjustment, just as the futile attempt to "save the family farm" of the last cycle was justified for a similar reason. But prolonging the end of the current cycle of consolidated power will be much more potentially dangerous because those being "helped" will not be family farmers but the wealthy classes who need no help. Expect the power structure to continue the present policy of giving welfare capital shamelessly to the rich in the name of saving society from starvation. Expect it to continue to perpetrate propaganda that hides the collapsing bonanza economy, especially the health hazards of factory food. Expect it to continue legal favoritism to the food monopolies.

In any event, there is one consolation for old agrarian cranks like me. Even under continued factory farm economics, monopoly can't work very well in food production as long as enough people have access to land, even backyard land, to grow food for themselves and local customers. We may not be able to make cars in our backyards cheaper than the moguls of money can do it, but give us land, any land, even the industrially bombed-out rubble of inner cities, and we can compete with factory food. Hardly anything is simpler or cheaper to do than raising chickens and vegetables.